Thermodynamic Weirdness

Thermodynamic Weirdness

From Fahrenheit to Clausius

Don S. Lemons

The MIT Press
Cambridge, Massachusetts
London, England

This book was set in ITC Stone Serif Std by Toppan Best-set Premedia Limited. Printed and bound in the United States of America.

Library of Congress Cataloging-in-Publication Data

Names: Lemons, Don S. (Don Stephen), 1949– author.
Title: Thermodynamic weirdness : from Fahrenheit to Clausius / Don S. Lemons.
Description: Cambridge, MA : The MIT Press, [2019] | Includes bibliographical references and index.
Identifiers: LCCN 2018020782 | ISBN 9780262039390 (hardcover : alk. paper)
Subjects: LCSH: Thermodynamics—History.
Classification: LCC TJ265 .L38375 2019 | DDC 536/.7—dc23 LC record available at https://lccn.loc.gov/2018020782

10 9 8 7 6 5 4 3 2 1

Contents

Preface

Physics, chemistry, and engineering students are taught the methods of classical thermodynamics, but its principles are not often scrutinized. As a result, heat is sometimes thought to be a form of motion, and the second law of thermodynamics is often, I maintain, misidentified as the law of increase of entropy.

These missteps are quite understandable. After all, those who write thermodynamics textbooks (a group that includes me) feel, above all else, the burden of preparing students to perform well on assigned problems. These books aim at analytical prowess and problem-solving skills—not at precise verbalizations of the laws and concepts of thermodynamics.

This problem-solving-first approach to thermodynamics works well for a certain group of students and teachers. After all, a verbal understanding of the laws and important concepts of classical thermodynamics can always be extracted from their mathematical formulation and application—if one takes the time to do so. And such is an excellent exercise—if attempted. Yet the conceptual structure of classical thermodynamics remains elusive for many, even for many of those who can apply its laws adeptly.

One difficulty is that thermodynamics emerged in the mid-nineteenth century not out of the earlier Newtonian synthesis but from a then more recent understanding of the limitations on heat flow and the newly discovered convertibility of work and heat. Sadi Carnot, Robert von Mayer, and James Joule established these understandings, while Rudolf Clausius and William Thomson (known late in life as Lord Kelvin) harmonized and expressed them in mathematical form. In this way, thermodynamics overthrew the hegemony of Newtonian concepts.

Because beginning physical science and engineering students, now as well as then, invest great effort in mastering Newtonian mechanics, they naturally attempt to incorporate all other subjects into its structure. Yet classical thermodynamics resists this incorporation. "Thermodynamic weirdness" predates "quantum weirdness" by at least a half century.[1]

One pedagogical response to thermodynamic weirdness is the creation of "thermal physics," an approach that interprets the laws and concepts of thermodynamics in terms of the Newtonian and quantum mechanics of particles as mediated through the statistics of large numbers, that is, through statistical mechanics. This approach is successful in many ways, but it has further marginalized the patterns of thought that brought thermodynamics into being.

Thermodynamic Weirdness responds to the problem-solving-first approach to thermodynamics and thermodynamic weirdness itself in another way: by focusing on the ideas of classical thermodynamics and their relationships, one to another, verbally

1. See D. S. L. Cardwell, *From Watt to Clausius: The Rise of Thermodynamics in the Early Industrial Age* (Ithaca, NY: Cornell University Press, 1971), 290–291.

expressed. Hence, this book avoids reference to developments that follow and add content to classical thermodynamics, in particular, the existence of atoms and molecules, statistical methods, and the ideas of quantum mechanics. Neither does the book develop mathematical methods. While this focus may limit the interest of some, I hope the resulting emphasis on concepts and intellectual structure attracts others. I have two kinds of readers in mind: teachers and students of thermodynamics who want to deepen their understanding of its classical formation and general readers who desire an introduction to its foundational ideas.

The history of thermodynamics is singularly helpful in this authorial and readerly task—more so, for instance, than, I think, the history of mechanics or the history of electromagnetism would be in parallel tasks. For this reason, the ideas in *Thermodynamic Weirdness* are chronologically as well as logically ordered, and primary sources that confirm and contextualize these ideas append most chapters. Included among the ideas explored and ordered are those, like the caloric or Thomson's 1848 definition of absolute temperature, that have in the fullness of time been found wanting, for the rise and fall of these ideas and others illumine, as no other light can, the more successful ideas that have taken their place.

Thermodynamic Weirdness also reflects my interest in the earliest versions of the second law of thermodynamics. While Carnot was the first, in 1824, to use a version of the second law, it was not until thirty years later, in 1854, that Clausius proved that the state variable we now call *entropy* must exist. Thus, postulating the second law of thermodynamics in terms of entropy, as is often done, obscures the development and possibly the meaning of these laws and concepts.

Even so, *Thermodynamic Weirdness* is not a history of thermodynamics. After all, I am a physicist and physics teacher, not a historian of science. A proper history of thermodynamics would do more to avoid "Whig history," a term of opprobrium invented by Herbert Butterfield to describe an approach to intellectual history in which ideas are either emphasized or ignored according to how well they remain current[2]—a sin of which the history of science is especially prone. Whig history allows the present to distort the past.

My aim in writing *Thermodynamic Weirdness* is, rather, to allow the past to enlighten the present and help readers appreciate the simplicity and coherence of classical thermodynamics, attractive features that have been obscured by its non-Newtonian origin and the current slant of pedagogical practice.

2. Herbert Butterfield, *The Whig Interpretation of History* (New York: Norton, 1965).

Acknowledgments

Thermodynamic Weirdness originated in my response to an anthology of primary sources on heat put together by Howard Fisher for his students at St. John's College, Santa Fe, New Mexico. Reading the words of Daniel Fahrenheit, Joseph Black, Antoine Lavoisier, Count Rumford, Sadi Carnot, Lord Kelvin, Robert Mayer, James Joule, and Rudolf Clausius gave me a new sense of their accomplishment and deepened my appreciation for and understanding of classical thermodynamics. These luminaries are still excellent teachers.

Howard Fisher also critically reviewed initial drafts of the text, as did Rick Shanahan, Joel Krehbiel, Galen Gisler, Reuben Hersh, and Harvey Leff. Kenneth Caneva offered a historian's perspective. Jesse Graber produced the figures. Blakely Mechau graciously translated, from the Latin, the excerpt from Daniel Fahrenheit's paper on the freezing of water.

Readers will notice the book's dependence on one frequently cited anthology of primary sources, *A Source Book in Physics* by W. F. Magie, and on two secondary sources: *Inventing Temperature: Measurement and Scientific Progress* by Hasok Chang and *From Watt to Clausius: The Rise of Thermodynamics in the Early Industrial*

Age by D. S. L. Cardwell. I acknowledge other literary debts in the annotated bibliography.

The staff at the MIT Press have, as usual, been quite helpful. Philip Laughlin, in particular, encouraged me to develop *Thermodynamic Weirdness* in certain directions and provided me with six insightful, anonymous referee reports. Of course, I could not follow all reviewer and referee suggestions, and some I have resisted. The text no doubt stubbornly reflects my own leanings. Even so, my aim in writing *Thermodynamic Weirdness* has been to present the ideas of classical thermodynamics as clearly and simply as possible.

Finally, my debts to fellow scholars and friends pale before those I owe my wife of many years, Allison Karslake Lemons. While not contributing directly to the content of *Thermodynamic Weirdness*, she has made its creation a joyful enterprise.

1
Inventing Temperature

1.1 Hot and Cold

Our sensitivity to hot and cold is a matter of life and death. Too hot or too cold, and we burn or freeze to death. We fashion clothes and build shelters to insulate us in some degree from these extremes.

Between too hot and too cold lie intermediate states, such as warm and cool, of which we have only a rough, idiosyncratic sense, for no two people feel degrees of temperature in exactly the same way. The different limbs of a single body do not always agree either, as attests the amusing demonstration of thrusting our two index fingers into a basin of tepid water after habituating one to cold and the other to hot water.

1.2 Thermometers

Apparently the first requirement for systematic thinking about degrees of hot and cold is to develop an inanimate but sensitive thermometer that always behaves in the same way under the

The title of this chapter is after the title of Hasok Chang's book *Inventing Temperature: Measurement and Scientific Progress* (Oxford: Oxford University Press, 2004).

same circumstances, a thermometer to which people in different conditions, places, and epochs might all have recourse.

Since most substances expand as their temperature (however felt) increases, the volume of a convenient thermometric substance serves this purpose. Why not correlate the volume of a certain quantity of water with the temperature of its surroundings? Marks on the side of a glass cylinder partially filled with water could indicate degrees of temperature. However, water, although readily available, is an inconvenient choice. After all, water may freeze in the winter, and when liquid water freezes, it expands rather than, as one might naively expect, contracts. In the eighteenth and early nineteenth centuries, mercury, alcohol, and air were all favored over water as a thermometric fluid.

Actually, a number of choices must be made in order to construct a thermometer. First, the thermometric fluid (mercury, alcohol, or air) must be chosen, then the thermometric variable (typically volume, but possibly pressure), and finally the fixed points (say, the freezing and boiling points of water) at which to specify the value of the temperature.

Furthermore, these choices are independent of the temperature scale that itself requires still more choices. Should the number of degrees of temperature between the freezing and boiling points of water be 80, 100, or 180? Should water freeze at 0 or at 32 degrees? And, as water becomes hotter, should its temperature decrease, as Anders Celsius (1701–1744) first proposed in 1742?

Daniel Fahrenheit (1686–1736) favored mercury as a thermometric fluid. After all, mercury has a freezing point well below and a boiling point much above those of water. He enclosed a column of mercury within a narrow, evacuated glass tube sealed at both ends. He associated evenly spaced marks on the glass with successive degrees of temperature. In 1724, Fahrenheit declared that water freezes at 32 degrees and that the normal temperature of the human body is 96 degrees. Fahrenheit was a professional glass blower, and the thermometers he made were small and rugged.

René Réaumur (1683–1757) favored alcohol as his thermometric fluid. He declared that the temperature of freezing water is 0 degrees and that every time the volume of alcohol expands by a factor of one one-thousandth, the temperature rises by 1 degree. In this way he found that water boils at 80 degrees—at least given the particular "spirit" or alcohol he used. (See figure 1.1 for a comparison of the Fahrenheit and Réaumur scales.)

Air thermometers were also possible, and although they are unwieldy, they had advantages that became apparent over time.

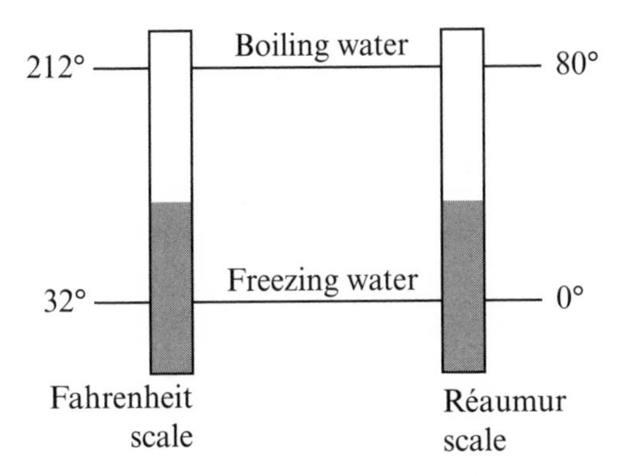

Figure 1.1
Fahrenheit and Réaumur scales.

1.3 Empirical Temperature

The temperature measured by a particular physically realized thermometer is called an *empirical temperature*. Of course, any given thermometer is composed of a particular thermometric fluid, with particular changes in the fluid correlated with particular intervals of temperature, and particular values of the temperature fixed at certain standard conditions.

Thus, any one state of affairs can have as many values of empirical temperature as there are kinds of thermometers measuring that state. In order to avoid confusion, one might, as a matter of convention, declare that one empirical thermometer is the standard against which all others should be calibrated much as today (in 2018) we calibrate all masses against the standard kilogram located in a vault in Saint-Cloud, France.

Yet why choose one or one kind of thermometer as standard? Would not the activities of scientists be just as well served by accompanying each temperature with a precise description of how that temperature was determined: the construction of the thermometer, the scale adopted, the way in which it was used, and so on? This view, sometimes called *operationalism*, defines the meaning of any measurement by the operations that produce it.

1.4 The Problem of Nomic Measurement

Not everyone was content with the arbitrariness of conventionalism or the agnosticism of operationalism. Indeed, most natural philosophers (as scientists were called until mid-nineteenth century) were realists, and most still are. In this context, realism means that when one measures a temperature, one believes one is measuring something more than the state of one's measuring instrument. Alternatively, the purpose of a thermometer is to produce a quantitative measure of an independent state of reality.

Consider, for instance, the problem of measuring the volume of a beaker of water. It is crucial that we define what the word *volume* means and that we believe that liquid water has a volume. This definition and this belief allow us to devise different methods of measuring the same volume. Our different measurements might be more or less precise, but we expect them to be consistent with one another.

Because they tell us something about that system's *state*, both the volume V and the temperature t are said to be *state variables*. (Here I adopt the tradition of representing empirical temperatures with a lowercase t and reserve the capital T for absolute temperatures as defined in chapters 4 and 8.) But there is an important difference between the two: volume can be measured directly, while temperature must be inferred from other measurements. In general, this inference depends on the correlation of the temperature t with other state variables that are directly measurable, such as the volume V or the pressure P of a thermometric fluid. But to establish this correlation, we must first know how to measure the temperature t. Hasok Chang neatly describes this circularity, also called the *problem of nomic* (that is, lawful) *measurement*, in the following way:[1]

1. We want to measure quantity X.
2. Because we cannot measure X directly, we must infer its value from another quantity Y that can be measured directly.

1. Chang, *Inventing Temperature*, 59. These four statements closely follow Chang's.

3. For this inference, we need a function $X = f(Y)$ that gives X in terms of Y.

4. Yet the form of this function cannot be known because that would require independently measuring values of X and Y that describe states of the system. And the value of X that describes a given state is exactly what we do not know and wish to find.

One might argue that by discovering a system's equation of state, we could use that system as a thermometer to avoid the problem of nomic measurement. Not so. Certainly Robert Boyle (1627–1691) discovered in 1662 that the pressure P of a given quantity of gas varies inversely with its volume V. And much later, in 1802, Joseph Louis Gay-Lussac (1778–1850) proposed that the pressure P of a gas is proportional to a certain linear function $(268+t)$ of its temperature t when the latter is denominated in "degrees centigrade."[2] Combining Boyle's and Gay-Lussac's laws, we have $PV = k \cdot (268+t)$ where k is a nonnegative, dimensional quantity that is proportional to the quantity of gas. Therefore,

$$t = \frac{PV}{k} - 268 \tag{1.1}$$

could be used to determine the temperature t of a gas in terms of its pressure P and volume V. The constant k and the number 268 could be determined by specifying the gas's temperature at two fixed points. According to equation (1.1), the lowest possible temperature is a –268 degrees centigrade.

But, of course, Gay-Lussac used some thermometer, possibly a mercury-in-glass thermometer, to establish the validity of

2. Today we would replace 268 in (268 + t) with 273 and also refer to "degrees Celsius."

equation (1.1) in order to determine t. For this reason, equation (1.1) might tell us nothing more than how the volume of the mercury in a mercury-in-glass thermometer correlates with the product of the pressure and volume of a certain quantity of gas.

1.5 Linearity, the Method of Mixtures, and Reproducibility

Eighteenth- and early nineteenth-century scientists searched for properties that distinguished good thermometers from bad ones. As we have seen, one property, assumed rather than discovered, was that the temperature should be a *linear* function of the thermometric variable. Thus, thermometer makers evenly spaced the degrees of temperature between the two fixed points—whatever the thermometric fluid adopted.

Yet different thermometers constructed on this principle disagreed with one another. Imagine, for instance, the unease of Herman Boerhaave (1668–1738) when he discovered in 1732 that two thermometers, one a mercury-in-glass and the other an alcohol thermometer, both constructed and calibrated with equally spaced degrees by Daniel Fahrenheit, produced different readings, except, of course, at the fixed points.[3] One of Fahrenheit's thermometers had to be wrong. But which one?

One way to answer this question lay in the "method of mixtures" according to which a mixture of two equal parts of water, one at boiling (100 degrees on the centigrade scale) and the other at freezing (0 degrees), should settle down at the average of the boiling and freezing temperatures (50 degrees). Similarly, combining two parts boiling and one part freezing water should

3. Chang, *Inventing Temperature*, 57–58.

produce water that is 67 degrees. A good thermometer, with its thermometric variable and temperature linearly related, should agree with these expectations.

The Swiss geologist and meteorologist Jean-André Deluc (1727–1817) published, in 1772, the result of his research comparing the readings of several thermometers composed of different thermometric fluids with those predicted by realizations of the method of mixtures. His conclusion: mercury alone satisfies the method of mixtures. "Certainly," declared Deluc, "nature gave us this mineral for making thermometers!"[4]

But Deluc's contemporaries had reasons to doubt this claim. One problem was that it seemed impossible to make mercury-in-glass thermometers with identical behavior. After all, glass was a significant part of a mercury-in-glass thermometer, and the composition and behavior of glass was then poorly understood. Today we know that glasses with different histories, even those of identical composition, can behave differently. In any case, identical composition could not be assumed. No wonder seemingly identical mercury-in-glass thermometers calibrated to agree at specific fixed points sometimes disagreed by several degrees elsewhere. In short, mercury-in-glass thermometers were *not reproducible.* And certainly reproducibility was as much a desired thermometric property as linearity and agreement with the method of mixtures.

1.6 Air Thermometers

Ultimately air replaced mercury as the most reliable thermometric fluid. Either the air's volume (with pressure constant)

4. Chang, *Inventing Temperature*, 64.

or its pressure (with volume constant) could serve as the thermometric variable. The temperature is always assumed to be a linear function of the thermometric variable. Also, air thermometers seemed to observe the requirements of the method of mixtures. And, importantly, air pressure increases much more rapidly with temperature than does the volume of a column of mercury. For this reason, the distorting effects of whatever contains the air are relatively small compared to the distorting effects of the glass of a mercury-in-glass thermometer. Thus, air thermometers are reproducible as, in the 1840s, the French experimentalist, Henri Victor Regnault (1810–1878), so meticulously demonstrated.[5]

Yet Regnault was careful not to conclude too much from his demonstration. After all, hydrogen gas and carbon dioxide also made good, reproducible thermometers, while the vapor of sulfuric acid did not. Why, and why not? Furthermore, given that the properties of water may not be uniform in temperature, why should we trust the method of mixtures? And, finally, why must the temperature be a *linear* function of the thermometric variable? These troubling questions remained unanswered until 1848 when William Thomson (1824–1907), later known as Lord Kelvin, discovered a definition of temperature that was independent of the behavior of any one particular substance.

Excerpts from primary sources follow most chapters. These excerpts highlight and contextualize important ideas and interpretations. My own comments are identified with my initials (DSL). Footnotes written by the original authors or editors are separately denoted from those in the main text.

5. Chang, *Inventing Temperature*, 83.

The following are the first two paragraphs of a paper by Daniel Fahrenheit on the freezing of water.—DSL

Daniel G. Fahrenheit, 1724

"Experimenta et Observationes de Congelatione aquae in vacuo factae" [Experiments and observations of the freezing of water done in a vacuum], *Philosophical Transactions* (London) 33, no. 782 (March–April 1724): 77. Translated from the Latin by Blakely Mechau.

Among the many wonderful phenomena of nature, I have always judged the freezing of water to be of no small importance; thus I have often been eager to test what might be the effect of cold if water were enclosed in a space without air. And since the second, third, and fourth days of March in the year 1721 looked favorably on this mode of experiment, I then made the following observations and experiments.

However, before I broach an account of the experiments, it will be necessary that I say a little something about the thermometers which are made by me, and about the division of their scale, and about the method I have used to evacuate them. Two kinds of thermometers are specially made by me, of which one is filled with alcohol and the other with mercury. Their lengths differ according to the use they are to serve. Yet all are in agreement because in all of them the division of their scale is the same; their differences relate to their fixed limits. The scale of the thermometers which are used only for meteorological observations begins at zero and ends at the 96th degree. The division of the scale of these thermometers depends on three fixed boundary points, which can be determined in the following manner; the first of those is found in the lowest part or, if you like, in the beginning of the scale. If a thermometer is placed in a mixture of ice, water, and ammonium chloride or even sea salt, its fluid descends to the degree which is denoted zero. This experiment succeeds better in

> winter than in summer. The second point is obtained if water and ice are mixed without the afore-mentioned salts. When the thermometer is placed in this mixture, its fluid reaches the 32nd degree. I call this point the beginning of freezing, for stagnant waters are already covered by the thinnest layer of ice when the fluid in the thermometer reaches this degree in winter. The third boundary point is found at the 96th degree. Alcohol extends as far as this degree when the thermometer is held for a long time in the mouth or in the armpit of a man in good health until it acquires most completely the heat of the body. But if the man's heat is feverish or he is suffering from some other heating disease, then it would have to be investigated by another thermometer the scale of which has been extended as far as the 128th or 132nd degree. I have not yet discovered by experiment whether these degrees would be sufficient for such a fever, but it is hardly to be believed that the degree of heat of any fever should exceed those mentioned. The scale of thermometers whose use is to investigate the degree of heat of boiling liquids still begins at zero and contains 600 degrees, for this is around the degree at which mercury itself (with which the thermometer is filled) begins to boil.

Here Fahrenheit seems to define a temperature scale with three fixed points: 0 degrees at which a mixture of ice, water, and ammonium chloride is stable; 32 degrees of a mixture of water and ice; and 96 degrees when the thermometer is placed in the mouth or under the armpit of a healthy man. But how can this be? We know that a linear scale, with its markings equally spaced on the glass of Fahrenheit's thermometer, requires exactly two fixed points and no more. That Fahrenheit did not adequately explain himself invites speculation.

Could it be that Fahrenheit defined two linear scales: one between the "fixed limits" or "boundary points" of the freezing of saltwater (0 degrees) and the freezing of relatively pure water (32 degrees) and a second between the freezing of pure water (32 degrees) and the *somewhat variable "blood heat" of a healthy person (96 degrees)? Of course, the first, lower scale could be extended to temperatures below 0 degrees and the second, higher scale to temperatures above 96 degrees. Possibly Fahrenheit did not recognize that, so defined, the size of a degree in the first interval, between 0 and 32 degrees, is not necessarily equal to the size of a degree in the second interval, between 32 and 96 degrees. Even so, the numerical convenience of the intervals between these fixed boundary points, separated as they are by successive powers of 2 ($32 = 2^5$ and $96 - 32 = 64 = 2^6$ degrees), may have been more important to Fahrenheit than continuous linearity.—DSL*

function of temperature. Yet accounting for this complication made the quantification and conservation of heat even more accurate.

Then in 1822, Joseph Fourier convincingly described heat as something that *diffuses* from one place to another rather than *flows* like an incompressible fluid. Heat diffuses through various materials much as a scent diffuses through the air, and as it diffuses, its density becomes more uniform. Yet even so, its total quantity is conserved. Indeed, Fourier built conservation into his mathematical description of the diffusion of heat and in this way created "the first branch of theoretical physics not to be based on Newton's laws of motion."[3]

2.2 Caloric

But what is heat? Two answers to this question remained current in the eighteenth century.[4] One was that heat is a form of motion, which is in some way communicated from hot bodies to cold ones, and another is that heat is a material fluid that flows from one object to another. Black, Lavoisier, and Laplace (during the latter's collaboration with Lavoisier) were cautious as to the ultimate nature of heat.[5] Their understanding of heat was primarily operational. Lavoisier even coined a word, *calorique*, to stand for "the matter of heat," that is, to stand for whatever causes the behavior of heat. According to Lavoisier, *calorique* "accords with every species of opinion."[6]

3. Cardwell, *From Watt to Clausius*, 190. Here Cardwell quotes a Dr. J. Herival.
4. A third theory of heat developed in the early nineteenth century: the "wave theory of heat." See Stephen G. Brush, *Statistical Physics and the Atomic Theory of Matter* (Princeton: Princeton University Press, 1983), 43–46.
5. Cardwell, *From Watt to Clausius*, 65.
6. See chapter 1 of Lavoisier's *Elements of Chemistry*.

Even so, the belief that heat is a material fluid that flows from one object to another was widely shared in the last quarter of the eighteenth century.[7] After all, the materiality of heat alone explained its conservation—and it was the conservation of heat that, in the near term, proved fruitful. According to the historian of thermodynamics D. S. L. Cardwell, "It is very difficult to see how a science of heat could have developed without a basic conservation principle."[8]

By the early nineteenth century, the materiality of heat had crystallized into the *doctrine of caloric*. Lavoisier's original caution was forgotten, and *calorique* came to stand for an ingenerate, indestructible "subtle fluid" composed of minute material particles that are repelled by one another and attracted to ordinary matter. For this reason, hot objects heat cold ones, and heated objects expand.

Eventually Laplace wholeheartedly adopted the doctrine of material caloric, and in the 1820s, he embarked on an ambitious program of research inspired by Newton's theory of gravitation. Laplace's research aimed at deriving the properties of heat from reasonable assumptions about the force among the supposed particles of caloric and between them and other kinds of particles.[9] While Laplace did have some success, he achieved nothing that others could not explain more simply. As a result, his contemporaries ignored his overly complex, speculative particle theory of caloric,[10] the failure of which underscored the non-Newtonian weirdness of heat.

7. Cardwell, *From Watt to Clausius*, 57.

8. Ibid., 27.

9. Hasok Chang, *Inventing Temperature: Measurement and Scientific Progress* (Oxford: Oxford University Press, 2004), 72–74.

10. C. C. Gillispie and R. L. Fox, *Pierre-Simon Laplace: 1749–1827: A Life in Exact Science* (Princeton: Princeton University Press, 1997), 248–249.

2.3 The Motion We Call Heat

Another answer to the question, "What is heat?" remained possible. As early as the mid-seventeenth century, Robert Boyle observed that continually striking a nail, already completely driven into a piece of wood, simply made the nail hotter. In this case, the motion of percussion causes heat. Friction does also. However, the explosive release of heat in a loaded cannon gives the cannonball a potentially destructive forward motion. In this instance, heat causes motion. If motion causes heat and heat causes motion, is not heat a form of motion?

One demonstration in favor of this *kinetic theory of heat* came from Count Rumford, known early in life as Benjamin Thompson (1753–1814). While making armaments for Prince-Elector Charles Theodore of Bavaria, Rumford enclosed, within a wooden box filled with water, a horse-powered, purposely dull drill bit fixed in brass cannon stock. After boring for two and a half hours, the initially cold water began to boil, much to the surprise of bystanders. Rumford concluded,

> It is hardly necessary to add that anything which any insulated body ... can continue to furnish without limitation cannot possibly be a material substance: and it appears to me to be extremely difficult, if not quite impossible, to form any distinct idea of anything capable of being excited and communicated, in the manner the heat was excited and communicated in these experiments, except it be motion. (1798)[11]

Rumford's demonstration has been much celebrated by those who later adopted the kinetic theory of heat. But for all its

11. Rumford, "An Inquiry Concerning the Source of Heat Which Is Excited by Friction," *Philosophical Transactions of the Royal Society of London* 88 (1798): 99. Also found in W. F. Magie, *A Source Book in Physics* (Cambridge, MA: Harvard University Press, 1935), 161.

drama, Rumford's demonstration did little to convince his contemporaries. Until John Dalton presented compelling chemical evidence for the existence of atoms in the early 1800s, the connection between heat and the motion of atoms was necessarily speculative. Even after Dalton, there was little agreement on how to quantify this "motion we call heat."[12] While suggestive, the kinetic theory of heat lacked predictive power.

2.4 *Heat* in Thermodynamics

That heat could be quantified and was, in many settings, conserved could not be doubted in the early nineteenth century even if its ultimate nature remained a mystery. Indeed, this calorimetric knowledge has a predictive power that today students of physical science master with ease. Simply mix so much of one material at one temperature with so much of another material at another temperature in a specially insulated container called a *calorimeter*, and the temperature to which this mixture approaches can, with the help of a table of specific and latent heats, be accurately predicted. Heat as a conserved quantity that diffuses from place to place was and is a highly successful idea.

For this reason, avoiding the word *heat* because of its presumed close connection with the now-discredited material theory of caloric leaves us with no convenient language with which to describe the operational rules of heat—in particular, its conservation and diffusion. Put more positively, what we can learn from the word *heat* (and from the word *caloric* as originally

12. After the title of an 1857 paper by Rudolf Clausius, "Über die Art der Bewegung, weiche wir Wärme nennen," *Annalen der Physik* 100 (1857): 353–380. English translation in *Philosophical Magazine*, series 4, 14 (1857): 108–127.

defined by Lavoisier) is the valuable distinction between behavior and an overarching theory that explains that behavior.

The natural philosophers of the early nineteenth century had no trouble with this distinction. They understood that the material theory of heat was simply one way of explaining the conservation of heat. For this reason, many of Rumford's contemporaries and successors remained agnostic as to the ultimate nature of heat.

For those desiring a definition of *heat* that is consistent with both what was, in the early nineteenth century, known of thermal phenomena and what today is known of thermodynamics, consider the following: *heat is that quantity which, in a system, is conserved in the absence of work done on or by that system.* In this way, the concept of heat as a conserved quantity is preserved in a particular limit (the limit of no work done) just as, for instance, Newtonian dynamics is preserved in a particular limit of special relativistic dynamics (the limit of small relative speeds).

Joseph Black is credited with identifying the phenomena of the latent heats of fusion and vaporization. Here he relates various measurements and observations in support of water having a significant heat of fusion. Otherwise, "melting snow would tear up and sweep away everything, and that so suddenly, that mankind should have great difficulty to escape from their ravages."—DSL

Joseph Black, 1764

Lectures on the Elements of Chemistry, compiled and published by John Robison (Philadelphia, 1807), 111–115.

Fluidity was universally considered as produced by a small addition to the quantity of heat which a body contains, when it is once heated up to its melting point; and the return of such body to a solid state as depending on a very small diminution of the quantity of its heat, after it is cooled to the same degree; that a solid body, when it is changed into a fluid, receives no greater addition to the heat within it than what is measured by the elevation of temperature indicated after fusion by the thermometer; and that, when the melted body is again made to congeal, by a diminution of its heat, it suffers no greater loss of heat than what is indicated also by the simple application to it of the same instrument.

This was the universal opinion on this subject, so far as I know, when I began to read my lectures in the University of Glasgow, in the year 1757. But I soon found reason to object to it, as inconsistent with many remarkable facts, when attentively considered; and I endeavored to show, that these facts are convincing proofs that fluidity is produced by heat in a very different manner.

I shall now describe the manner in which fluidity appeared to me to be produced by heat, and we shall then compare the former and my view of the subject with the phenomena.

The opinion I formed from attentive observation of the facts and phenomena is as follows. When ice, for example, or any other solid substance, is changing into a fluid by heat, I am of opinion that it receives a much greater quantity of heat than what is perceptible in it immediately after by the thermometer. A great quantity of heat enters into it, on this occasion, without making it apparently warmer, when tried by that instrument. This heat, however, must be thrown into it, in order to give it the form of a fluid; and I affirm, that this great addition of heat is the principal, and most immediate cause of the fluidity induced.

And, on the other hand, when we deprive such a body of its fluidity again, by a diminution of its heat, a very great quantity of heat comes out of it, while it is assuming a solid form, the loss of which heat is not to be perceived by the common manner of using the thermometer. The apparent heat of the body, as measured by that instrument, is not diminished, or not in proportion to the loss of heat which the body actually gives out on this occasion; and it appears from a number of facts, that the state of solidity cannot be induced without the abstraction of this great quantity of heat. And this confirms the opinion, that this quantity of heat, absorbed, and, as it were, concealed in the composition of fluids, is the most necessary and immediate cause of their fluidity.

To perceive the foundation of this opinion, and the inconsistency of the former with many obvious facts, we must consider, in the first place, the appearances observable in the melting of ice, and the freezing of water.

If we attend to the manner in which ice and snow melt, when exposed to the air of a warm room, or, when a thaw succeeds to frost, we can easily perceive, that however cold they might be at the first, they are soon heated up to their melting point, or begin soon at their surface to be changed into water. And if the common opinion had been well founded, if the complete change of them into water required only the further addition of a very small quantity of heat, the mass, though of a considerable size, ought all to be melted in a very few minutes or seconds more, the heat continuing incessantly to be communicated from the air around. Were this really the case, the consequences of it would be dreadful, in many cases; for, even as things are at present, the melting of great quantities of snow and ice occasions violent torrents, and great inundations in the cold countries, or in the rivers that come from them. But, were the ice and snow to melt as suddenly as they must necessarily do, were the former opinion of the action of heat in melting them well founded, the torrents and inundations would be incomparably more irresistible and dreadful. They would tear up and sweep away everything, and that so suddenly, that mankind should have great difficulty to escape from their ravages. This sudden liquefaction does not actually happen; the masses of ice or snow melt with a very slow progress, and require a long time, especially if they be of a large size, such as are the collections of ice, and wreaths of snow, formed in some places during the winter. These, after they begin to melt, often require many weeks of warm weather, before they are totally dissolved into, water. This remarkable slowness with which ice is melted, enables us to preserve it easily during the summer, in the structures called ice-houses. It begins to melt in these, as soon as it is put into them; but, as the building exposes only a small surface to the air, and has a very thick covering of thatch, and the access of the external air to the inside of it is prevented as much as possible, the heat penetrates the ice-house with a slow progress, and this, added to the slowness with which the ice itself is disposed to melt, protracts the total liquefaction of it so long, that some of it remains to the end of summer. In the same manner does snow continue on many mountains during the whole summer, in a melting state, but melting so slowly, that the whole of that season is not a sufficient time for its complete liquefaction.

This remarkable slowness with which ice and snow melt, struck me as quite inconsistent with the common opinion of the modification of heat, in the liquefaction of bodies.

And this very phenomenon is partly the foundation of the opinion I have proposed; for if we examine what happens, we may perceive that a great quantity of heat enters the melting ice, to form the water into which it is changed, and that the length of time necessary for the collection of so much heat from the surrounding bodies, is the reason of the slowness with which the ice is liquefied. If any person entertains doubts of the entrance and absorption of heat in the melting ice, he needs only to touch it; he will instantly feel that it rapidly draws heat from his warm hand. He may also examine the bodies that surround it, or are in contact with it, all of which he will find deprived by it of a great part of their heat; or if he suspend it by a thread, in the air of a warm room, he may perceive with his hand, or by a thermometer, a stream of cold air descending constantly from the ice; for the air in contact is deprived of a part of its heat, and thereby condensed and made heavier than the warmer air of the rest of the room; it therefore falls downwards, and its place round the ice is immediately supplied by some of the warmer air; but this, in its turn, is soon deprived of some heat, and prepared to descend in like manner; and thus there is a constant flow of warm air from around, to the sides of the ice, and a descent of the same in a cold state, from the lower part of the mass, during which operation the ice must necessarily receive a great quantity of heat.

It is, therefore, evident, that the melting ice receives heat very fast, but the only effect of this heat is to change it into water, which is not in the least sensibly warmer than the ice was before. A thermometer, applied to the drops or small streams of water, immediately as it comes from the melting ice, will point to the same degree as when it is applied to the ice itself, or if there is any difference, it is too small to deserve notice. A great quantity, therefore, of the heat, or of the matter of heat, which enters into the melting ice, produces no other effect but to give it fluidity, without augmenting its sensible heat; it appears to be absorbed and concealed within the water, so as not to be discoverable by the application of a thermometer.

Lavoisier's first task in his Elements of Chemistry *was to discuss the language of chemistry. In the course of this discussion, Lavoisier coined a new word,* caloric, *to stand for the "matter of heat." One purpose of this coinage was to "accord with every species of opinion; since, strictly speaking, we are not obliged to suppose this to be a real substance." If Lavoisier often wrote of caloric as if it were a real or material substance, he also warned that "it is especially necessary to guard against the extravagancy of our imagination, which forever inclines to step beyond the bounds of truth, and is with difficultly restrained within the narrow limits of facts."—DSL*

Antoine Lavoisier, 1789

Elements of Chemistry, 5th ed. (Edinburgh, 1802), 1:49–56. Translated from the French by Robert Kerr.

Part I, Chapter I. Of the Combinations of Caloric and the Formation of Elastic Aeriform Fluids

That every body, whether solid or fluid, is augmented in all its dimensions by any increase of its sensible heat, was long ago fully established as a physical axiom, or universal proposition, by the celebrated Boerhaave. Such facts as have been adduced for controverting the generality of this principle, offer only fallacious results, or, at least, such as are so complicated with foreign circumstances, as to mislead the judgment. But, when we separately consider the effects, so as to deduce each from the cause to which they separately belong, it is easy to perceive, that the separation of particles by heat is a constant and general law of nature.

When we have heated a solid body to a certain degree, and have thereby caused its particles to separate from each other, if we allow the body to cool, its particles again approach each other, in the same proportion in which they were separated by the increased temperature; the body returns by the same degrees of expansion through which it before extended; and, if brought back to the same temperature which it possessed at the commencement of the experiment, it recovers exactly the same dimensions which it formerly occupied. We are still very far from being able to produce the degree of absolute cold, or total deprivation of heat, being unacquainted with any degree of coldness which we cannot suppose capable of still further augmentation; hence it follows, that we are incapable of causing the ultimate particles of bodies to approach each other as near as possible, and that these particles of bodies do not touch each other in any state hitherto known. Though this be a very singular conclusion, it is impossible to be denied.

It may be supposed, that, since the particles of bodies are thus continually impelled by heat to separate from each other, they would have no connection between themselves; and that, of consequence, there could be no solidity in nature, unless these particles were held together by some other power which tended to unite them, and, so to speak, to chain them together: This power, whatever be its cause, or manner of operation, is named Attraction.

Thus the particles of all bodies may be considered as subject to the action of two opposite powers, Repulsion and Attraction, between which they remain in equilibrium. So long as the attractive force remains stronger, the body must continue in a state of solidity; but if, on the contrary, heat has so far removed these particles from each other as to place them beyond the sphere of attraction, they lose the cohesion they before had with each other, and the body ceases to be solid.

Water gives us a regular and constant example of these facts. While its temperature is below 32° of Fahrenheit's scale* it remains solid, and is called ice. Above that degree of temperature, its particles being no longer held together by reciprocal attraction, it becomes liquid; and, when we raise its temperature above 212°, its particles, giving way to the repulsion caused by the heat, assume the state of vapor or gas, and the water is changed into an aeriform fluid.

The same may be affirmed of all bodies in nature. They are either solid, or liquid, or in the state of elastic aeriform vapor, according to the proportion which takes place between the attractive force inherent in their particles, and the repulsive power of the heat acting upon these; or, what amounts to the same thing, in proportion to the degrees of heat to which they are exposed.

It is difficult to comprehend these phenomena, without admitting them as the effects of a real and material substance, or very subtle fluid, which, insinuating itself between the particles of bodies, separates them from each other. Even allowing that the existence of this fluid may be hypothetical, we shall see in the sequel that it explains the phenomena of nature in a very satisfactory manner.

*Whenever the degree of heat occurs in the original, it is stated by the author according to Réaumur's thermometer; but the translator has thought it more convenient to use Fahrenheit's scale, as more generally employed and understood in Britain.—Translator

This substance, whatever it is, being the cause of heat, or, in other words, the sensation which we call *warmth* being caused by the accumulation of this substance, we cannot, in strict language, distinguish it by the term *heat*, because the same name would then very improperly express both cause and effect. For this reason, in the memoir which I published in 1777,[†] I gave it the names of *igneous fluid* and *matter of heat*: And, since that time, in the work[‡] published by Mr de Morveau, Mr Berthollet, Mr de Fourcroy, and myself, upon the reformation of chemical nomenclature, we thought it necessary to reject all periphrastic expressions, which both lengthen physical language, and render it less distinct, and which even frequently do not convey sufficiently just ideas of the object intended. Wherefore, we have distinguished the cause of heat, or that exquisitely elastic fluid which produces it, by the term of *caloric*. Besides that this expression fulfills our object in the system which we have adopted, it possesses this farther advantage, that it accords with every species of opinion; since, strictly speaking, we are not obliged to suppose this to be a real substance, it being sufficient, as will more clearly appear in the sequel of this work, that it be considered as the repulsive stuff, whatever that may be, which separates the particles of matter from each other; so that we are still at liberty to investigate its effects in an abstract and mathematical manner.

In the present state of our knowledge, we are unable to determine whether light be a modification of caloric, or if caloric be, on the contrary a modification of light. This, however, is indisputable, that in a system where only decided facts are admissible, and where we avoid, as far as possible, to suppose any thing to be, that is not really known to exist, we ought provisionally to distinguish, by distinct terms, such things as are known to produce different effects. We therefore distinguish light from caloric; though we do not therefore deny that these have certain qualities in common, and that, in certain circumstances, they combine with other

[†]Collections of the French Academy of Sciences for that year, 420.
[‡]*New Chemical Nomenclature.*

bodies almost in the same manner, and produce, in part, the same effects.

What I have already said may suffice to determine the idea affixed to the word *caloric*; but there remains a more difficult attempt, which is, to give a just conception of the manner in which caloric acts upon other bodies. Since this subtle matter penetrates through the pores of all known substances; since there are no vessels through which it cannot escape; and, consequently, as there are none which are capable of retaining it; we can only come at the knowledge of its properties by effects which are fleeting and difficultly ascertainable. It is in those things, which we neither see nor feel, that it is especially necessary to guard against the extravagancy of our imagination, which forever inclines to step beyond the bounds of truth, and is with difficultly restrained within the narrow limits of facts.

Count Rumford's most valuable contribution to classical thermodynamics was to identify a phenomenon, the seemingly unlimited production of heat during the boring of cannon, that was at variance with the prevailing ideology of the conservation of heat. In place of conservation, Rumford proposed the vague notion that heat is motion. Yet to Rumford, heat remains "one of those mysteries of nature which are beyond the reach of human intelligence."

Although born in Woburn, Massachusetts, Benjamin Thompson (1753–1814), as Rumford was then known, allied himself with the British during the American War of Independence, after which he left America and moved to England. In 1785 Thompson departed England for Bavaria, where he served the prince-elector for eleven years. For his inventions and designs, the prince made him Count Rumford of the Holy Roman Empire in 1791.—DSL

Benjamin Thompson (Count Rumford), 1798

"An Inquiry Concerning the Source of the Heat Which Is Excited by Friction," in Benjamin Rumford, G. E. Ellis, and American Academy of Arts and Sciences, *The Complete Works of Count Rumford* (London: Macmillan, 1876), vol. 1, 471–493.

It frequently happens that in the ordinary affairs and occupations of life, opportunities present themselves of contemplating some of the most curious operations of Nature; and very interesting philosophical experiments might often be made, almost without trouble or expense, by means of machinery contrived for the mere mechanical purposes of the arts and manufactures.

I have frequently had occasion to make this observation; and am persuaded that a habit of keeping the eyes open to everything that is going on in the ordinary course of the business of life has oftener led, as it were by accident, or in the playful excursions of the imagination, put into action by contemplating the most common appearances, to useful doubts and sensible schemes for investigation and improvement, than all the more intense meditations of philosophers in the hours expressly set apart for study.

It was by accident that I was led to make the experiments of which I am about to give an account; and, though they are not perhaps of sufficient importance to merit so formal an introduction, I cannot help flattering myself that they will be thought curious in several respects, and worthy of the honor of being made known to the Royal Society.

Being engaged lately in superintending the boring of cannon in the workshops of the military arsenal at Munich, I was struck with the very considerable degree of Heat which a brass gun acquires in a short time in being bored, and with the still more intense Heat (much greater than that of boiling water, as I found by experiment) of the metallic chips separated from it by the borer.

The more I meditated on these phenomena, the more they appeared to me to be curious and interesting. A thorough investigation of them seemed even to bid fair to give a farther insight into the hidden nature of Heat; and to enable us to form some reasonable conjectures respecting the existence, or non-existence, of an igneous fluid, a subject on which the opinions of philosophers have in all ages been much divided.

In order that the Society may have clear and distinct ideas of the speculations and reasonings to which these appearances gave rise in my mind, and also of the specific objects of philosophical investigation they suggested to me, I must beg leave to state them at some length, and in such manner as I shall think best suited to answer this purpose.

Rumford arranged for two horses to turn a partially hollowed-out, horizontally mounted cylinder of brass cannon stock, 7.75 inches in diameter and 9.8 inches long, while a stationary, blunted drill bit was pressed along the axis of and against the solid interior part of the cylinder with approximately a thousand pounds of force. A small hole had been drilled perpendicular to the axis of the solid part of the brass cylinder in order to periodically receive a thermometer. Furthermore, "To prevent, as far as possible, the loss of any part of the Heat that was generated in the experiment, the cylinder was well covered up with a fit coating of thick and warm flannel, which was carefully wrapped round it, and defended it on every side from the cold air of the atmosphere."

Rumford described several kinds of measurements with this arrangement that eliminated the possibility that merely separating pieces of brass from the cannon stock releases latent heat stored within the cannon stock, that the specific heat capacity of the pieces of brass was different from that of the stock, and that the surrounding air played an essential role in the generation of heat.

For his experiments 3 and 4, the description of which follows, Rumford constructed a watertight wooden box that surrounded the cannon stock and the dull drill bit.—DSL

Everything being ready, I proceeded to make the experiment I had projected in the following manner.

The hollow cylinder having been previously cleaned out, and the inside of its bore wiped with a clean towel till it was quite dry, the square iron bar, with the blunt steel borer fixed to the end of it, was put into its place; the mouth of the bore of the cylinder being closed at the same time by means of the circular piston, through the center of which the iron bar passed.

This being done, the box was put in its place, and the joinings of the iron rod and of the neck of the cylinder with the two ends of the box having been made water tight by means of collars of oiled leather, the box was filled with cold water (viz. at the temperature of 60°), and the machine was put in motion.

The result of this beautiful experiment was very striking, and the pleasure it afforded me amply repaid me for all the trouble I had had in contriving and arranging the complicated machinery used in making it.

The cylinder, revolving at the rate of about 32 times in a minute, had been in motion but a short time, when I perceived, by putting my hand into the water and touching the outside of the cylinder, that Heat was generated; and it was not long before the water which surrounded the cylinder began to be sensibly warm.

At the end of 1 hour I found, by plunging a thermometer into the water in the box (the quantity of which fluid amounted to 18.77 lb, avoirdupois, or 2 and 1/4 wine gallons), that its temperature had been raised no less than 47 degrees; being now 107° of Fahrenheit's scale.

When 30 minutes more had elapsed, or 1 hour and 30 minutes after the machinery had been put in motion, the Heat of the water in the box was 142°.

At the end of 2 hours, reckoning from the beginning of the experiment, the temperature of the water was found to be raised to 178°.

At 2 hours 20 minutes it was at 200°; and at 2 hours 30 minutes it actually boiled!

It would be difficult to describe the surprise and astonishment expressed in the countenances of the by standers, on seeing so large a quantity of cold water heated, and actually made to boil, without any fire.

Though there was, in fact, nothing that could justly be considered as surprising in this event, yet I acknowledge fairly that it afforded me a degree of childish pleasure, which, were I ambitious of the reputation of a grave philosopher, I ought most certainly rather to hide than to discover. ...

By meditating on the results of all these experiments, we are naturally brought to that great question which has so often been the subject of speculation among philosophers; namely, What is Heat? Is there any such thing as an igneous fluid? Is there anything that can with propriety be called caloric?

We have seen that a very considerable quantity of Heat may be excited in the friction of two metallic surfaces, and given off in a constant stream or flux in all directions without interruption or intermission, and without any signs of diminution or exhaustion.

From whence came the Heat which was continually given off in this manner in the foregoing experiments? Was it furnished by the small particles of metal, detached from the larger solid masses, on their being rubbed together? This, as we have already seen, could not possibly have been the case.

Was it furnished by the air? This could not have been the case; for, in three of the experiments, the machinery being kept immersed in water, the access of the air of the atmosphere was completely prevented.

Was it furnished by the water which surrounded the machinery? That this could not have been the case is evident: first, because this water was continually receiving Heat from the machinery, and could not at the same time be giving to, and receiving Heat from, the same body; and, secondly, because there was no chemical decomposition of any part of this water. Had any such decomposition taken place (which, indeed, could not reasonably have been expected), one of its component elastic fluids (most probably inflammable air) must at the same time have been set at liberty, and, in making its escape into the atmosphere, would have been detected; but though I frequently examined the water to see if any air-bubbles rose up through it, and had even made preparations for catching them, in order to examine them, if any should appear, I could perceive none; nor was there any sign of decomposition of any kind whatever, or other chemical process, going on in the water.

Is it possible that the Heat could have been supplied by means of the iron bar to the end of which the blunt steel borer was fixed or by the small neck of gun-metal by which the hollow cylinder was united to the cannon? These suppositions appear more improbable even than either of those before mentioned; for Heat was continually going off, or out of the machinery, by both these passages, during the whole time the experiment lasted.

And, in reasoning on this subject, we must not forget to consider that most remarkable circumstance, that the source of the Heat generated by friction, in these experiments, appeared evidently to be inexhaustible.

It is hardly necessary to add, that anything which any insulated body, or system of bodies, can continue to furnish without limitation, cannot possibly be a material substance; and it appears to me to be extremely difficult, if not quite impossible, to form any distinct idea of any thing capable of being excited and communicated in the manner the Heat was excited and communicated in these experiments, except it be motion.

I am very far from pretending to know how, or by what means or mechanical contrivance, that particular kind of motion in bodies which has been supposed to constitute Heat is excited, continued, and propagated; and I shall not presume to trouble the Society with mere conjectures, particularly on a subject which, during so many thousand years, the most enlightened philosophers have endeavored, but in vain, to comprehend.

But, although the mechanism of Heat should, in fact, be one of those mysteries of nature which are beyond the reach of human intelligence, this ought by no means to discourage us or even lessen our ardor, in our attempts to investigate the laws of its operations. How far can we advance in any of the paths which science has opened to us before we find ourselves enveloped in those thick mists which on every side bound the horizon of the human intellect? But how ample and how interesting is the field that is given us to explore!

Nobody, surely, in his sober senses, has ever pretended to understand the mechanism of gravitation; and yet what sublime discoveries was our immortal Newton enabled to make, merely by the investigation of the laws of its action!

The effects produced in the world by the agency of Heat are probably just as extensive, and quite as important, as those which are owing to the tendency of the particles of matter towards each other; and there is no doubt but its operations are, in all cases, determined by laws equally immutable.

Before I finish this Essay, I would beg leave to observe, that although, in treating the subject I have endeavored to investigate, I have made no mention of the names of those who have gone over the same ground before me, nor of the success of their labors, this omission has not been owing to any want of respect for my predecessors, but was merely to avoid prolixity, and to be more at liberty to pursue, without interruption, the natural train of my own ideas.

3
Carnot's Analysis

3.1 Steam Engines

The first practical steam engines, built from Thomas Newcomen's design of 1712, consumed wood or coal and produced heat that boiled water and made its vapor push against a movable piston head. The motion of this piston, when linked to a pump, pulled water from the bottom of English coal, iron, copper, and tin mines. Eventually steam engines, as improved by James Watt (1736–1819), not only drained mines but also powered cotton mills, drove rail locomotives, and, consequently, brought the industrial revolution to England and the rest of the world.

The head of the piston in a Newcomen engine was pushed out and in—out as steam filled the piston chamber and pushed against the piston head during its *expansion stroke* and in as cold water, injected into the piston chamber, condensed the steam and produced a partial vacuum that allowed the atmosphere to push the piston head back during its *compression stroke*. As a consequence, steam and cold water alternately heated and cooled the piston.

Watt, starting in 1765, improved on Newcomen's design: after its expansion stroke, the steam vented into a separate chamber, the *condenser*, into which the cold water was then injected. In this way, the piston could remain hot and the condenser cold—a feature that, for reasons then unknown, increased the "duty" of a steam engine.[1]

3.2 Carnot's Law

The steam engines that empowered and enriched England were only one kind of a class of devices called *heat engines*: machines that produce motion or work from heat. Heat engines puzzled the young French military engineer Sadi Carnot (1796–1832). Carnot's teachers and fellow students at the École Polytechnique (among them Pierre-Simon Laplace, André-Marie Ampère, Jean-Baptist Biot, Francois Arago, and Augustin-Jean Fresnel), in the early nineteenth century were busy constructing detailed mathematical theories, in particular, of electromagnetic and optical phenomena. And mechanical processes were everywhere explained in terms of Newtonian pushes and pulls. Heat engines alone resisted analysis. Carnot alone, among his generation, rose to this challenge.

Unfortunately, Carnot published only once before his death at age thirty-six. The aim of his *Reflections on the Motive Power of Fire* (1824) was to explain heat engines from a "general point of view ... independently of any mechanism or agent ... applicable to all heat engines ... whatever the working substance and whatever the method by which it is operated."[2]

1. The *duty* of a steam engine is work produced per unit fuel consumed. The modern equivalent of duty is *efficiency*.
2. Sadi Carnot, *Reflections on the Motive Power of Fire* (Gloucester, MA: Peter Smith, 1977) 6.

Early in the *Reflections* Carnot directed the reader's attention to an important circumstance:

> The production of motion in steam engines is always accompanied by a circumstance on which we should fix our attention. This circumstance is ... [heat's] passage from a body in which temperature is more or less elevated to another in which it is lower. (1824)

Carnot found this "circumstance on which we should fix our attention" so important he repeated its point in each of seven consecutive paragraphs in the first few pages of the *Reflections*. Examples of these repetitions are: "[the passage of caloric] from a more or less heated body to a cooler one,"[3] "its transportation from a warm body to a cold body," and "These changes are not caused by uniform temperature, but rather by alternations of hot and cold." He also remarked on the "self-evident" fact that "wherever there exists a difference of temperature, motive power can be produced."[4]

This self-evident fact, which I call *Carnot's law*, is most conveniently expressed in terms of *reservoirs*, that is, in terms of indefinitely large, constant-temperature bodies that serve as sources or sinks of heat. In this way, Carnot's law becomes, *No heat engine can produce work without transporting heat from a reservoir at a higher temperature to one at a lower temperature.*[5]

3. Carnot used the words *heat* and *caloric* interchangeably. See his footnote on p. 9 of *Reflections*.

4. Carnot, *Reflections*, 6–9. Note that for Carnot, *temperature* meant *empirical temperature*.

5. See section 5.2 of D. S. Lemons, *Mere Thermodynamics* (Baltimore: Johns Hopkins University Press, 2009), and D. S. Lemons and M. K. Penner, "Sadi Carnot's Contribution to the Second Law of Thermodynamics," *American Journal of Physics* 76 (2008): 21–25.

3.3 Carnot's Waterwheel

In the midst of his analysis Carnot compared a heat engine to a waterwheel. (See the overshot waterwheel in figure 3.1.) While not necessary to the logic of the *Reflections*, Carnot's waterwheel neatly illustrates what he claimed of heat engines: they require both high- and low-temperature reservoirs and conserve heat as they produce work. A waterwheel and a heat engine both produce "motive power" by virtue of something falling without diminution, water or heat, in height or temperature, from a higher to a lower level.

This comparison may have been inspired by the work of Sadi's father, Lazare Carnot, a leader of the French Revolution and able

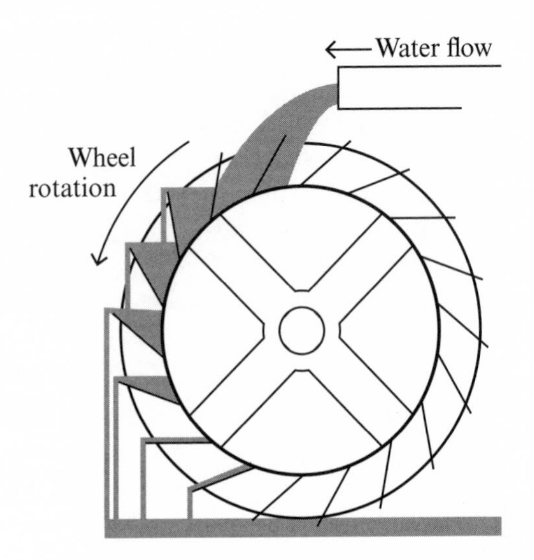

Figure 3.1
Overshot waterwheel.

mathematician and scientist who educated his son until age sixteen. Lazare had written a treatise that Sadi might have studied, *Fundamental Principles of Equilibrium and Movement*, on the efficiency of mechanical devices composed of wheels, pulleys, inclined planes, and levers. Lazare demonstrated in his text that these devices deliver the most work when they avoid shocks and accelerations and, of course, friction in their moving parts.[6]

Sadi might also have learned from his father's treatise that every instance of water falling any distance without steadily pushing on the wheel is a waste of resource. Such unexploited falls should be minimized, if not altogether eliminated. The analogous waste in a heat engine occurs when heat "falls" from a hot body to a cold one without producing work. For this reason, the hot and cold parts of a heat engine should never touch. Watt's introduction of a separate condensing chamber was a step in this direction: it increased the duty of his steam engine, for then the piston was no longer periodically heated by the entering steam and cooled by the injected cold water.

3.4 Carnot's Engine

Carnot's quest to describe a heat engine in general led inevitably to the concept of a *perfect* heat engine—or, as we now call it, a *Carnot engine*. To adapt a statement from Tolstoy, all perfect heat engines are alike; each imperfect heat engine is imperfect in its own way.[7] Carnot came to believe that a perfect heat engine should transport a quantity of heat, without diminution, from a

6. E. Mendoza, introduction to *Reflections on the Motive Power of Heat* (Gloucester, MA: Peter Smith, 1977), x.

7. The first sentence of Leo Tolstoy's *Anna Karenina* (New York: Penguin, 2000) is "All happy families are alike; each unhappy family is unhappy in its own way."

hotter to a colder reservoir; that the parts of a perfect heat engine should move without friction, acceleration, or deceleration; that its hot and cold parts should never touch; and that its working fluid should suffer no turbulence or dissipation.

Carnot provided a design in which these features were present—abstracted from inessentials such as boilers, condensers, and mechanical linkages. Even steam is unnecessary. Never mind that no one can build a perfect heat engine. Knowing its design might help engineers approach this ideal in some degree.

The *working fluid* of a Carnot engine is any substance that expands when heated and contracts when cooled. This fluid is sealed within the piston chamber and is initially in thermal contact with and at the same temperature as the hotter heat reservoir. The piston then cycles through the four motions illustrated in figure 3.2. First, as the fluid expands and pushes out the piston head, the fluid temperature falls imperceptibly below that of the hotter reservoir with which it is in thermal contact and so absorbs heat from it. This motion, and similar ones, should be

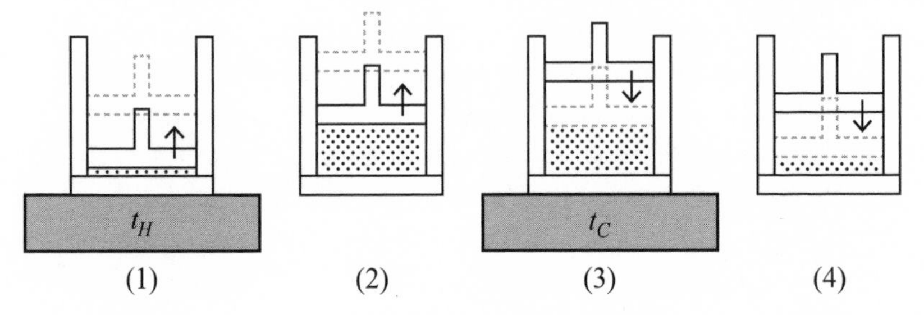

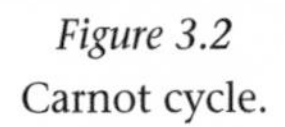
Figure 3.2
Carnot cycle.

indefinitely slow, that is, *quasi-static,* in order to keep the fluid temperature indefinitely close to that of the reservoir with which it is in thermal contact and to avoid initiating fluid turbulence and dissipation. Second, the piston is removed from the hotter heat reservoir and thermally insulated. The fluid expands quasi-statically and pushes out the piston head until the fluid temperature drops to that of the colder reservoir. Third, the piston is placed in thermal contact with the colder reservoir while its head quasi-statically compresses the fluid. In this way, the fluid transfers heat to the colder reservoir while remaining in thermal contact with it. Finally, the piston is removed from the colder reservoir and thermally insulated. The piston head quasi-statically compresses the fluid until its temperature rises to that of the hotter reservoir. The fluid is then returned to thermal contact with the hotter reservoir, and the cycle repeats.

In step 1, the fluid quasi-statically expands while in thermal contact with the hotter reservoir; in step 2, the fluid quasi-statically expands while thermally insulated; in step 3, the fluid is quasi-statically compressed while in thermal contact with the colder reservoir; and in step 4, the fluid is quasi-statically compressed while thermally isolated.[8]

This cycle of motions, a *Carnot cycle,* produces in its expansion stroke (steps 1 and 2) more work than is consumed in its compression stroke (steps 3 and 4). After all, the temperature of the fluid is hotter in the former than in the latter, and the fluid pushes more forcefully on the piston head when hot and

8. The processes in steps 1 and 3 are called *isothermal,* that is, same-temperature, because in them, the temperature of the working fluid departs only negligibly from that of the constant-temperature heat reservoir with which it is in thermal contact. The processes in steps 2 and 4 are called *adiabatic,* that is, incapable of being crossed, because in them, the working fluid is insulated.

expanding than resists compression when cold and contracting. Yet observe that because the Carnot cycle proceeds quasi-statically, that is, indefinitely slowly, the rate at which it produces work, that is, its *power* output, is vanishingly small. For this reason, the Carnot cycle is more a conceptual tool than the design of a practical engine.

3.5 Reversibility and Carnot's Theorem

Because the steps of a Carnot cycle are executed quasi-statically and without friction and dissipation and without hot and cold parts touching, their motions are *reversible*; the steps of a Carnot engine can be performed in reverse order and direction with only indefinitely small changes in the fluid or its environment. When the four steps of a Carnot cycle are reversed, the piston compresses the working fluid when the latter is hot and expands it when cold. In this way, the Carnot engine consumes rather than produces work and absorbs heat from the cold reservoir and transfers it to the hot one—essentially what a modern refrigerator does.[9]

Carnot used the reversibility of a Carnot engine to prove an important theorem, one we now call *Carnot's theorem*. According to Carnot's theorem, *no heat engine operating in a cycle between two heat reservoirs produces more work per unit heat transported than a reversible one*. If an irreversible heat engine could produce more work per unit heat transported than a reversible one, then only part of the work produced by this hypothetical, superproductive, irreversible engine would be needed to run the Carnot engine in

9. Simple refrigerators were constructed in the eighteenth century. But it was James Harrison, a British journalist and inventor, who in 1856 built the first commercially practical refrigerators after having immigrated to Australia.

reverse until the latter restores to the hotter reservoir the heat transported to the colder reservoir by the irreversible engine. In this way, the two engines working together, one reversible and one superproductive and irreversible, could produce net work with no net transport of heat, that is, in Carnot's phrase, could produce "unlimited ... motive power without consumption either of caloric or of any other agent whatever"—surely an impossibility.

Figure 3.3 illustrates this proof by contradiction. The engine in panel a is reversible and has, in fact, in the left-hand part of panel c, been reversed. By supposition, the superproductive, irreversible engine of panel b absorbs the same heat Q from the same hotter reservoir as that of panel a and rejects the same heat Q to the same cooler reservoir as does that of panel a but produces more work—$W + \Delta$ instead of W. In panel c, the two engines have been combined. The irreversible part of the engine provides the reversed part with just enough of the work it produces W to transport heat Q from the colder to the hotter reservoir. Therefore, the two engines working together produce net work Δ with no net transport of heat. In sum, the result of assuming the contrary (a superproductive, irreversible heat engine exists) leads to an absurdity (work can be produced with no net transport of heat or consumption of fuel) that requires the assumption to be rejected.

3.6 Carnot's Function

Two other propositions follow immediately from the definition of a Carnot cycle and the concept of reversibility: (1) *All reversible heat engines operating cyclically between the same two heat reservoirs produce the same work per unit heat W / Q*, and (2) *the work per unit*

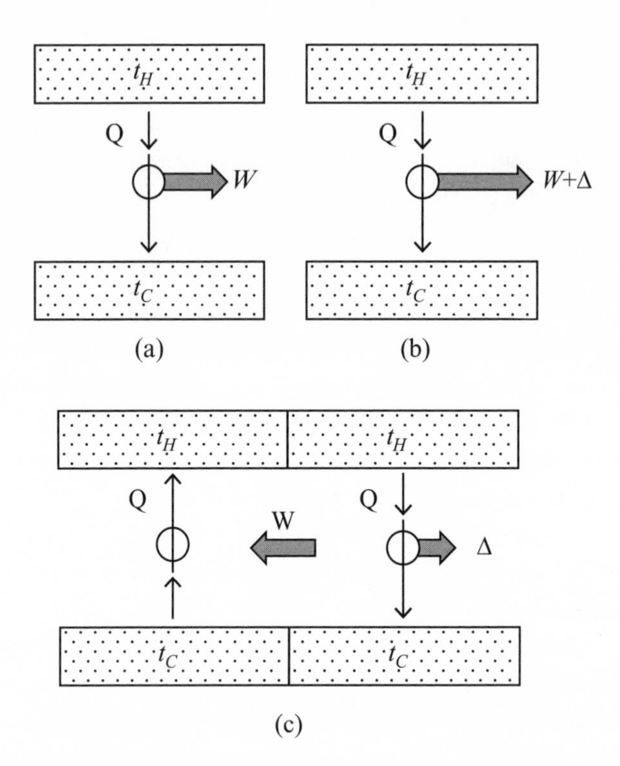

Figure 3.3
(a) Reversible, cyclic heat engine. (b) Hypothetical, superproductive, irreversible heat engine. (c) Reversed heat engine with input from the hypothetical, superproductive, irreversible engine results in net work Δ produced with no net transport of heat.

heat produced W/Q by a reversible engine operating cyclically between two heat reservoirs is a function $\varepsilon(t_H, t_C)$ of only the empirical temperatures t_H and t_C of the two reservoirs, that is,

$$\frac{W}{Q} = \varepsilon(t_H, t_C). \tag{3.1}$$

The diagrams of figure 3.3, with slight reinterpretation, illustrate indirect proofs (that is, proofs by contradiction) of these theorems. In each case the absurdity produced is *perpetual motion,* that is, the production of work with no net heat transported—or, more plainly, with no consumption of fuel.

Carnot tried, without success, to discover the form of what we now call *Carnot's function,* $\varepsilon(t_H, t_C)$. Given what Carnot knew when writing the *Reflections* in 1824, one can demonstrate only that $\varepsilon(t_H, t_C)$ decreases as t_C approaches t_H from below and that $\varepsilon(t_H, t_C)$ vanishes altogether when $t_C = t_H$.[10]

It would be gratifying to report that others took up the challenge of discovering the form of Carnot's function. Émile Clapeyron (1799–1864), in a memoir published in 1834, did translate many of Carnot's results into mathematical and graphical form. But otherwise, Carnot's reflections were ignored for almost twenty-five years. Only in 1848 did William Thomson (1824–1907) discover and make significant use of Clapeyron's memoir and in this way give Carnot's analysis the notice it deserved.

Those of Sadi Carnot's arguments that are most determinative of the structure of classical thermodynamics appear in the Reflections *in*

10. Lemons, *Mere Thermodynamics,* chap. 5.

verbal form. These describe a perfect heat engine—limited only by the nature of things, in particular, by Carnot's Law or what is now known as the second law of thermodynamics. The whole document occupies some ninety pages. This approximately twenty-page selection conveys Carnot's primary concerns and outlines his main arguments.—DSL

Sadi Carnot, 1824

*Reflections on the Motive Power of Heat and on Machines Fitted to Develop That Power** (New York: Wiley, 1897), 37–61. Translated from the French by R. H. Thurston, 2nd rev. ed.

Every one knows that heat can produce motion. That it possesses vast motive-power no one can doubt, in these days when the steam engine is everywhere so well known.

To heat also are due the vast movements which take place on the earth. It causes the agitations of the atmosphere, the ascension of clouds, the fall of rain and of meteors, the currents of water which channel the surface of the globe, and of which man has thus far employed but a small portion. Even earthquakes and volcanic eruptions are the result of heat.

From this immense reservoir we may draw the moving force necessary for our purposes. Nature, in providing us with combustibles on all sides, has given us the power to produce, at all times and in all places, heat and the impelling power which is the result of it. To develop this power, to appropriate it to our uses, is the object of heat engines.

*Sadi Carnot's *Reflexions sur la puissance motrice du feu* (Paris: Bachelier, 1824) was long ago completely exhausted. As but a small number of copies were printed, this remarkable work remained long unknown to the earlier writers on Thermodynamics. It was therefore for the benefit of savants unable to study a work out of print, as well as to render honor to the memory of Sadi Carnot, that the new publishers of the *Annales Scientifique de École Normale supérieure* (i. series, 1.1, 1872) published a new edition, from which this translation is reproduced.—Translator.

The study of these engines is of the greatest interest, their importance is enormous, their use is continually increasing, and they seem destined to produce a great revolution in the civilized world.

Already the steam-engine works our mines, impels our ships, excavates our ports and our rivers, forges iron, fashions wood, grinds grains, spins and weaves our cloths, transports the heaviest burdens, etc. It appears that it must some day serve as a universal motor, and be substituted for animal power, waterfalls, and air currents. Over the first of these motors it has the advantage of economy, over the two others the inestimable advantage that it can be used at all times and places without interruption.

If, some day, the steam engine shall be so perfected that it can be set up and supplied with fuel at small cost, it will combine all desirable qualities, and will afford to the industrial arts a range the extent of which can scarcely be predicted. It is not merely that a powerful and convenient motor that can be procured and carried anywhere is substituted for the motors already in use, but that it causes rapid extension in the arts in which it is applied, and can even create entirely new arts.

The most signal service that the steam engine has rendered to England is undoubtedly the revival of the working of the coal mines, which had declined, and threatened to cease entirely, in consequence of the continually increasing difficulty of drainage, and of raising the coal.[†] We should rank second the benefit to iron manufacture, both by the abundant supply of coal substituted for wood just when the latter had begun to grow scarce, and by the powerful machines of all kinds, the use of which the introduction of the steam engine has permitted or facilitated.

[†]It may be said that coal mining has increased tenfold in England since the invention of the steam engine. It is almost equally true in regard to the mining of copper, tin, and iron. The results produced in a half-century by the steam engine in the mines of England are today paralleled in the gold and silver mines of the New World—mines of which the working declined from day to day, principally on account of the insufficiency of the motors employed in the draining and the extraction of the minerals.

Iron and heat are, as we know, the supporters, the bases, of the mechanic arts. It is doubtful if there be in England a single industrial establishment of which the existence does not depend on the use of these agents, and which does not freely employ them. To take away today from England her steam engines would be to take away at the same time her coal and iron. It would be to dry up all her sources of wealth, to ruin all on which her prosperity depends, in short, to annihilate that colossal power. The destruction of her navy, which she considers her strongest defense, would perhaps be less fatal.

The safe and rapid navigation by steamships may be regarded as an entirely new art due to the steam engine. Already this art has permitted the establishment of prompt and regular communications across the arms of the sea, and on the great rivers of the old and new continents. It has made it possible to traverse savage regions where before we could scarcely penetrate. It has enabled us to carry the fruits of civilization over portions of the globe where they would else have been wanting for years. Steam navigation brings nearer together the most distant nations. It tends to unite the nations of the earth as inhabitants of one country. In fact, to lessen the time, the fatigues, the uncertainties, and the dangers of travel—is not this the same as greatly to shorten distances?[‡]

The discovery of the steam engine owed its birth, like most human inventions, to rude attempts which have been attributed to different persons, while the real author is not certainly known. It is, however, less in the first attempts that the principal discovery consists, than in the successive improvements which have brought steam engines to the condition in which we find them today. There is almost as great a distance between the first apparatus in which the expansive force of steam was displayed and the

[‡]We say, to lessen the dangers of journeys. In fact, although the use of the steam-engine on ships is attended by some danger which has been greatly exaggerated, this is more than compensated by the power of following always an appointed and well known route, of resisting the force of the winds which would drive the ship toward the shore, the shoals, or the rocks.

existing machine, as between the first raft that man ever made and the modern vessel.

If the honor of a discovery belongs to the nation in which it has acquired its growth and all its developments, this honor cannot be here refused to England. Savery, Newcomen, Smeaton, the famous Watt, Woolf, Trevithick, and some other English engineers, are the veritable creators of the steam engine. It has acquired at their hands all its successive degrees of improvement. Finally, it is natural that an invention should have its birth and especially be developed, be perfected, in that place where its want is most strongly felt.

Notwithstanding the work of all kinds done by steam engines, notwithstanding the satisfactory condition to which they have been brought today, their theory is very little understood, and the attempts to improve them are still directed almost by chance.

The question has often been raised whether the motive power of heat* is unbounded, whether the possible improvements in steam engines have an assignable limit,—a limit which the nature of things will not allow to be passed by any means whatever; or whether, on the contrary, these improvements may be carried on indefinitely. We have long sought, and are seeking today, to ascertain whether there are in existence agents preferable to the vapor of water for developing the motive power of heat; whether atmospheric air, for example, would not present in this respect great advantages. We propose now to submit these questions to a deliberate examination.

The phenomenon of the production of motion by heat has not been considered from a sufficiently general point of view. We have considered it only in machines the nature and mode of action of which have not allowed us to take in the whole extent of application of which it is susceptible. In such machines the phe-

*We use here the expression motive power to express the useful effect that a motor is capable of producing. This effect can always be likened to the elevation of a weight to a certain height. It has, as we know, as a measure, the product of the weight multiplied by the height to which it is raised.

nomenon is, in a way, incomplete. It becomes difficult to recognize its principles and study its laws.

In order to consider in the most general way the principle of the production of motion by heat, it must be considered independently of any mechanism or any particular agent. It is necessary to establish principles applicable not only to steam-engines[†] but to all imaginable heat engines, whatever the working substance and whatever the method by which it is operated.

Machines which do not receive their motion from heat, those which have for a motor the force of men or of animals, a waterfall, an air current, etc., can be studied even to their smallest details by the mechanical theory. All cases are foreseen, all imaginable movements are referred to these general principles, firmly established, and applicable under all circumstances. This is the character of a complete theory. A similar theory is evidently needed for heat engines. We shall have it only when the laws of Physics shall be extended enough, generalized enough, to make known beforehand all the effects of heat acting in a determined manner on any body.

We will suppose in what follows at least a superficial knowledge of the different parts which compose an ordinary steam engine; and we consider it unnecessary to explain what are the furnace, boiler, steam cylinder, piston, condenser, etc.

The production of motion in steam engines is always accompanied by a circumstance on which we should fix our attention. This circumstance is the re-establishing of equilibrium in the caloric; that is, its passage from a body in which the temperature is more or less elevated, to another in which it is lower. What happens in fact in a steam engine actually in motion? The caloric developed in the furnace by the effect of the combustion traverses the walls of the boiler, produces steam, and in some way incorporates itself with it. The latter carrying it away, takes it first into the

[†]We distinguish here the steam engine from the heat engine in general. The latter may make use of any agent whatever, of the vapor of water or of any other, to develop the motive power of heat.

cylinder, where it performs some function, and from thence into the condenser, where it is liquefied by contact with the cold water which it encounters there. Then, as a final result, the cold water of the condenser takes possession of the caloric developed by the combustion. It is heated by the intervention of the steam as if it had been placed directly over the furnace. The steam is here only a means of transporting the caloric. It fills the same office as in the heating of baths by steam, except that in this case its motion is rendered useful.

We easily recognize in the operations that we have just described the reestablishment of equilibrium in the caloric, its passage from a more or less heated body to a cooler one. The first of these bodies, in this case, is the heated air of the furnace; the second is the condensing water. The reestablishment of equilibrium of the caloric takes place between them, if not completely, at least partially, for on the one hand the heated air, after having performed its function, having passed round the boiler, goes out through the chimney with a temperature much below that which it had acquired as the effect of combustion; and on the other hand, the water of the condenser, after having liquefied the steam, leaves the machine with a temperature higher than that with which it entered.

The production of motive power is then due in steam engines not to an actual consumption of caloric, but *to its transportation from a warm body to a cold body,* that is, to its reestablishment of equilibrium—an equilibrium considered as destroyed by any cause whatever, by chemical action such as combustion, or by any other. We shall see shortly that this principle is applicable to any machine set in motion by heat.

According to this principle, the production of heat alone is not sufficient to give birth to the impelling power: it is necessary that there should also be cold; without it, the heat would be useless. And in fact, if we should find about us only bodies as hot as our furnaces, how can we condense steam? What should we do with it if once produced? We should not presume that we might discharge it into the atmosphere, as is done in some engines;[‡] the atmosphere would not receive it. It does receive it under the actual condition of things, only because it fulfills the office of a vast condenser, because it is at a lower temperature; otherwise it would soon become fully charged, or rather would be already saturated.*

Wherever there exists a difference of temperature, wherever it has been possible for the equilibrium of the caloric to be reestablished, it is possible to have also the production of impelling power. Steam is a means of realizing this power, but it is not the only one. All substances in nature can be employed for this purpose, all are susceptible of changes of volume, of successive contractions and dilatations, through the alternation of heat and cold. All are capable of overcoming in their changes of volume certain resistances, and of thus developing the impelling power. A solid body—a metallic bar for example—alternately heated and cooled increases and diminishes in length, and can move bodies fastened to its ends. A liquid alternately heated and cooled increases and diminishes in volume, and can overcome obstacles of greater or less size, opposed to its dilatation. An aeriform fluid is susceptible

[‡]Certain engines at high pressure throw the steam out into the atmosphere instead of the condenser. They are used specially in places where it would be difficult to procure a stream of cold water sufficient to produce condensation.

*The existence of water in the liquid state here necessarily assumed, since without it the steam-engine could not be fed, supposes the existence of a pressure capable of preventing this water from vaporizing, consequently of a pressure equal or superior to the tension of vapor at that temperature. If such a pressure were not exerted by the atmospheric air, there would be instantly produced a quantity of steam sufficient to give rise to that tension, and it would be necessary always to overcome this pressure in order to throw out the steam from the engines into the new atmosphere. Now this is evidently equivalent to overcoming the tension which the steam retains after its condensation, as effected by ordinary means.

If a very high temperature existed at the surface of our globe, as it seems certain that it exists in its interior, all the waters of the ocean would be in a state of vapor in the atmosphere, and no portion of it would be found in a liquid state.

of considerable change of volume by variations of temperature. If it is enclosed in an expansible space, such as a cylinder provided with a piston, it will produce movements of great extent. Vapors of all substances capable of passing into a gaseous condition, as of alcohol, of mercury, of sulphur, etc., may fulfill the same office as vapor of water. The latter, alternately heated and cooled, would produce motive power in the shape of permanent gases, that is, without ever returning to a liquid state. Most of these substances have been proposed, many even have been tried, although up to this time perhaps without remarkable success.

We have shown that in steam engines the motive power is due to a re-establishment of equilibrium in the caloric; this takes place not only for steam engines, but also for every heat-engine—that is, for every machine of which caloric is the motor. Heat can evidently be a cause of motion only by virtue of the changes of volume or of form which it produces in bodies.

These changes are not caused by uniform temperature, but rather by alternations of heat and cold. Now to heat any substance whatever requires a body warmer than the one to be heated; to cool it requires a cooler body. We supply caloric to the first of these bodies that we may transmit it to the second by means of the intermediary substance. This is to re-establish, or at least to endeavor to re-establish, the equilibrium of the caloric.

It is natural to ask here this curious and important question: Is the motive power of heat invariable in quantity, or does it vary with the agent employed to realize it as the intermediary substance, selected as the subject of action of the heat?

It is clear that this question can be asked only in regard to a given quantity of caloric,[†] the difference of the temperatures also

[†]It is considered unnecessary to explain here what is quantity of caloric or quantity of heat (for we employ these two expressions indifferently), or to describe how we measure these quantities by the calorimeter. Nor will we explain what is meant by latent heat, degree of temperature, specific heat, etc. The reader should be familiarized with these terms through the study of the elementary treatises of physics or of chemistry.

being given. We take, for example, one body A kept at a temperature of 100° and another body B kept at a temperature of 0°, and ask what quantity of motive power can be produced by the passage of a given portion of caloric (for example, as much as is necessary to melt a kilogram of ice) from the first of these bodies to the second. We inquire whether this quantity of motive power is necessarily limited, whether it varies with the substance employed to realize it, whether the vapor of water offers in this respect more or less advantage than the vapor of alcohol, of mercury, a permanent gas, or any other substance. We will try to answer these questions, availing ourselves of ideas already established.

We have already remarked upon this self-evident fact, or fact, which at least appears evident as soon as we reflect on the changes of volume occasioned by heat: *wherever there exists a difference of temperature, motive power can be produced.* Reciprocally, wherever we can consume this power, it is possible to produce a difference of temperature; it is possible to occasion destruction of equilibrium in the caloric. Are not percussion and the friction of bodies actually means of raising their temperature, of making it reach spontaneously a higher degree than that of the surrounding bodies, and consequently of producing a destruction of equilibrium in the caloric, where equilibrium previously existed? It is a fact proved by experience, that the temperature of gaseous fluids is raised by compression and lowered by rarefaction. This is a sure method of changing the temperature of bodies, and destroying the equilibrium of the caloric as many times as may be desired with the same substance. The vapor of water employed in an inverse manner to that in which it is used in steam engines can also be regarded as a means of destroying the equilibrium of the caloric. To be convinced of this we need but to observe closely the manner in which motive power is developed by the action of heat on vapor of water. Imagine two bodies A and B, kept each at a constant temperature, that of A being higher than that of B. These two bodies, to which we can give or from which we can remove the heat without causing their temperatures to vary, exercise the functions of two unlimited reservoirs of caloric. We will call the first the furnace and the second the refrigerator.

If we wish to produce motive power by carrying a certain quantity of heat from the body A to the body B we shall proceed as follows:

(1) To borrow caloric from the body A to make steam with it—that is, to make this body fulfill the function of a furnace, or rather of the metal composing the boiler in ordinary engines—we here assume that the steam is produced at the same temperature as the body A.

(2) The steam having been received in a space capable of expansion, such as a cylinder furnished with a piston, to increase the volume of this space, and consequently also that of the steam. Thus rarefied, the temperature will fall spontaneously, as occurs with all elastic fluids; admit that the rarefaction may be continued to the point where the temperature becomes precisely that of the body B.

(3) To condense the steam by putting it in contact with the body B, and at the same time exerting on it a constant pressure until it is entirely liquefied. The body B fills here the place of the injection-water in ordinary engines, with this difference, that it condenses the vapor without mingling with it, and without changing its own temperature.[‡]

[‡]We may perhaps wonder here that the body B being at the same temperature as the steam is able to condense it. Doubtless this is not strictly possible, but the slightest difference of temperature will determine the condensation, which suffices to establish the justice of our reasoning. It is thus that, in the differential calculus, it is sufficient that we can conceive the neglected quantities indefinitely reducible in proportion to the quantities retained in the equations, to make certain of the exact result.

The body B condenses the steam without changing its own temperature—this results from our supposition. We have admitted that this body may be maintained at a constant temperature. We take away the caloric as the steam furnishes it. This is the condition in which the metal of the condenser is found when the liquefaction of the steam is accomplished by

The operations which we have just described might have been performed in an inverse direction and order. There is nothing to prevent forming vapor with the caloric of the body B, and at the temperature of that body, compressing it in such a way as to make it acquire the temperature of the body A, finally condensing it by contact with this latter body, and continuing the compression to complete liquefaction.

By our first operations there would have been at the same time production of motive power and transfer of caloric from the body A to the body B. By the inverse operations there is at the same time expenditure of motive power and return of caloric from the body B to the body A. But if we have acted in each case on the same quantity of vapor, if there is produced no loss either of motive power or caloric, the quantity of motive power produced in the first place will be equal to that which would have been expended in the second, and the quantity of caloric passed in the first case from the body A to the body B would be equal to the quantity which passes back again in the second from the body B to the body A; so that an indefinite number of alternative operations of this sort could be carried on without in the end having either produced motive power or transferred caloric from one body to the other.

Now if there existed any means of using heat preferable to those which we have employed, that is, if it were possible by any method whatever to make the caloric produce a quantity of motive power greater than we have made it produce by our first series of operations, it would suffice to divert a portion of this power in order by the method just indicated to make the caloric of the

---

applying cold water externally, as was formerly done in several engines. Similarly, the water of a reservoir can be maintained at a constant level if the liquid flows out at one side as it flows in at the other.

One could even conceive the bodies A and B maintaining the same temperature, although they might lose or gain certain quantities of heat. If, for example, the body A were a mass of steam ready to become liquid, and the body B a mass of ice ready to melt, these bodies might, as we know, furnish or receive caloric without thermometric, change.

body B return to the body A from the refrigerator to the furnace, to restore the initial conditions, and thus to be ready to commence again an operation precisely similar to the former, and so on: this would be not only perpetual motion, but an unlimited creation of motive power without consumption either of caloric or of any other agent whatever. Such a creation is entirely contrary to ideas now accepted, to the laws of mechanics and of sound physics. It is inadmissible.* We should then conclude that *the maximum of motive power resulting from the employment of steam is also the maximum of motive power realizable by any means whatever.* We will soon give a second more rigorous demonstration of this theory. This should be considered only as an approximation.

*The objection may perhaps be raised here, that perpetual motion, demonstrated to be impossible by mechanical action alone, may possibly not be so if the power either of heat or electricity be exerted; but is it possible to conceive the phenomena of heat and electricity as due to anything else than some kind of motion of the body, and as such should they not be subjected to the general laws of mechanics? Do we not know besides, a posteriori, that all the attempts made to produce perpetual motion by any means whatever have been fruitless?—that we have never succeeded in producing a motion veritably perpetual, that is, a motion which will continue forever without alteration in the bodies set to work to accomplish it? The electromotor apparatus (the pile of Volta) has sometimes been regarded as capable of producing perpetual motion; attempts have been made to realize this idea by constructing dry piles said to be unchangeable; but however it has been done, the apparatus has always exhibited sensible deteriorations when its action has been sustained for a time with any energy. The general and philosophic acceptation of the words perpetual motion should include not only a motion susceptible of indefinitely continuing itself after a first impulse received, but the action of an apparatus, of any construction whatever, capable of creating motive power in unlimited quantity, capable of starting from rest all the bodies of nature if they should be found in that condition, of overcoming their inertia; capable, finally, of finding in itself the forces necessary to move the whole universe, to prolong, to accelerate incessantly, its motion. Such would be a veritable creation of motive power. If this were a possibility, it would be useless to seek in currents of air and water or in combustibles this motive power. We should have at our disposal an inexhaustible source upon which we could draw at will.

Carnot does indeed give a more precise version of this argument in a part of the text not excerpted here.—DSL

We have a right to ask, in regard to the proposition just enunciated, the following questions: What is the sense of the word maximum here? By what sign can it be known that this maximum is attained? By what sign can it be known whether the steam is employed to greatest possible advantage in the production of motive power?

Since every reestablishment of equilibrium in the caloric may be the cause of the production of motive power, every reestablishment of equilibrium which shall be accomplished without production of this power should be considered as an actual loss. Now, very little reflection would show that all change of temperature which is not due to a change of volume of the bodies can be only a useless re-establishment of equilibrium in the caloric.[†] The necessary condition of the maximum is, then, that in the bodies employed to realize the motive power of heat there should not occur any change of temperature which may not be due to a change of volume. Reciprocally, every time that this condition is fulfilled the maximum will be attained. This principle should never be lost sight of in the construction of heat engines; it is its fundamental basis. If it cannot be strictly observed, it should at least be departed from as little as possible.

Every change of temperature which is not due to a change of volume or to chemical action (an action that we provisionally suppose not to occur here) is necessarily due to the direct passage of the caloric from a more or less heated body to a colder body.

[†]We assume here no chemical action between the bodies employed to realize the motive power of heat. The chemical action which takes place in the furnace is, in some sort, a preliminary action—an operation destined not to produce immediately motive power, but to destroy the equilibrium of the caloric, to produce a difference of temperature which may finally give rise to motion.

This passage occurs mainly by the contact of bodies of different temperatures; hence such contact should be avoided as much as possible. It cannot probably be avoided entirely, but it should at least be so managed that the bodies brought in contact with each other differ as little as possible in temperature. When we just now supposed, in our demonstration, the caloric of the body A employed to form steam, this steam was considered as generated at the temperature of the body A; thus the contact took place only between bodies of equal temperatures; the change of temperature occurring afterwards in the steam was due to dilatation, consequently to a change of volume. Finally, condensation took place also without contact of bodies of different temperatures. It occurred while exerting a constant pressure on the steam brought in contact with the body B of the same temperature as itself. The conditions for a maximum are thus found to be fulfilled. In reality the operation cannot proceed exactly as we have assumed. To determine the passage of caloric from one body to another, it is necessary that there should be an excess of temperature in the first, but this excess may be supposed as slight as we please. We can regard it as insensible in theory, without thereby destroying the exactness of the arguments. ...

According to established principles at the present time, we can compare with sufficient accuracy the motive power of heat to that of a waterfall. Each has a maximum that we cannot exceed, whatever may be, on the one hand, the machine which is acted upon by the water, and whatever, on the other hand, the substance acted upon by the heat. The motive power of a waterfall depends on its height and on the quantity of the liquid; the motive power of heat depends also on the quantity of caloric used, and on what may be termed, on what in fact we will call, the height of its fall,[‡] that is to say, the difference of temperature of the bodies between which the exchange of caloric is made. In the waterfall the motive

[‡]The matter here dealt with being entirely new, we are obliged to employ expressions not in use as yet, and which perhaps are less clear than is desirable.

power is exactly proportional to the difference of level between the higher and lower reservoirs. In the fall of caloric the motive power undoubtedly increases with the difference of temperature between the warm and the cold bodies; but we do not know whether it is proportional to this difference. We do not know, for example, whether the fall of caloric from 100 to 50 degrees furnishes more or less motive power than the fall of this same caloric from 50 to zero. It is a question which we propose to examine hereafter.

4
Absolute Temperature

4.1 William Thomson

The year 1845 was full of promise for the young William Thomson (1824–1907), for he had just become a Cambridge fellow and was about to assume the chair of natural philosophy at the University of Glasgow.[1] But before taking up his professorial duties, Thomson visited Victor Regnault's laboratory in Paris. Regnault (1810–1878) tutored Thomson in the difficulties of thermometry and the inherently empirical and seemingly arbitrary nature of temperature measurements.

While in Paris, Thomson read Émile Clapeyron's 1834 paper on the motive power of heat and noted his use of Carnot's *Reflections*.[2] Thus began Thomson's fruitful encounter with Carnot's ideas. Initially Thomson had to depend on Clapeyron's account of the latter, for the Paris booksellers he approached in 1845 had

1. Thomson is better known as Lord Kelvin, a title he adopted when elevated to the nobility in 1892, or simply as Kelvin.
2. Emile Clapeyron, "Memoir on the Motive Power of Heat," trans. E. Mendoza in *Reflections on the Motive Power of Fire and Other Papers on the Second Law of Thermodynamics* (Gloucester, MA: Peter Smith, 1977, 1960), 71–105.

never heard of Sadi Carnot. Not until 1849 did Thomson obtain a copy of the *Reflections*.[3]

4.2 Thomson's 1848 Definition of Absolute Temperature

In the meantime, Thomson discovered in Carnot's analysis a way to establish an *absolute temperature scale* independent of any one thermometric fluid. In place of mercury, alcohol, or air, Thomson proposed using the hypothetical thermometric fluid sealed within the piston of a Carnot engine—by design, bereft of any particular properties except those of expanding when heated and contracting when cooled. In place of equal increments of volume marked on the side of a glass tube, Thomson proposed using equal increments of work produced by identical reversible heat engines connected in series between two heat reservoirs. By definition, these heat engines operate between heat reservoirs separated by equal degrees of absolute temperature.[4]

Figure 4.1 illustrates Thomson's definition. One heat reservoir is at the temperature T_H, say, of boiling water and the other at the temperature T_C, say, of freezing water. (Recall that here and elsewhere, an uppercase T represents an absolute temperature, and a lowercase t stands for an empirical temperature.) Suppose a Carnot engine, that is, a reversible, cyclic engine,

3. D. S. L. Cardwell, *From Watt to Clausius: The Rise of Thermodynamics in the Early Industrial Age* (Ithaca, NY: Cornell University Press, 1971), 240.
4. William Thomson, "On an Absolute Thermometric Scale, Founded on Carnot's Theory of the Motive Power of Heat, and Calculated from Regnault's Observations," *Philosophical Magazine* (October 1848) [from Sir William Thomson, *Mathematical and Philosophical Papers*, vol. 1 (Cambridge: Cambridge University Press, 1882), 100–106]. Thomson's paper also appears in W. F. Magie, *A Source Book in Physics* (Cambridge, MA: Harvard University Press, 1963, 1935), 237–242.

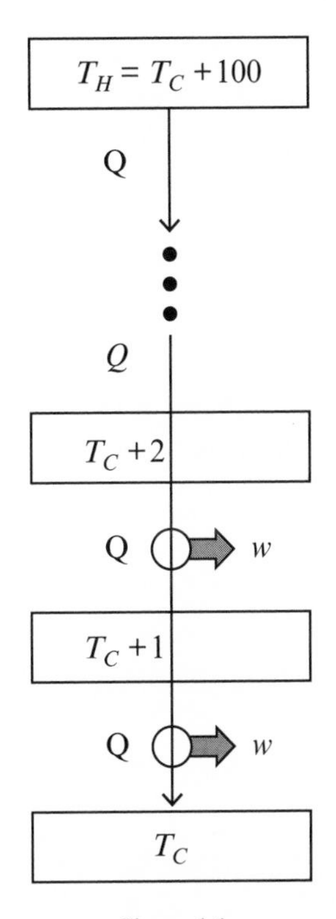

Figure 4.1
Carnot engines, connected in series, that illustrate Thomson's 1848 definition of absolute temperature.

operating between these two heat reservoirs produces work W while transporting a quantity of heat Q from the hotter to the colder reservoir. Then divide this Carnot engine into 100 identical ones connected in series with each producing the same work $w[= W/100]$ and passing on the same heat Q. The nth Carnot engine in this series operates between a hotter reservoir at temperature T_n and a cooler one at temperature T_{n-1} whose difference $T_n - T_{n-1}$ is defined so that $T_n - T_{n-1} = 1$.

This was Thomson's first definition of absolute temperature. (Several years later, he proposed a second, ultimately more successful, one.) He summarized his idea as follows:

> The characteristic property of the scale I now propose is, that all degrees have the same value [of work produced]; that is, that a unit of heat descending from body A at the temperature $T°$ of the scale, to a body B at the temperature $(T-1)°$, would give out the same mechanical effect, whatever be the number T. This may justly be termed an absolute scale, since its characteristic is quite independent of the physical properties of any specific substance. (1848)

There is nothing illogical or incoherent in Thomson's 1848 definition of absolute temperature as long as one assumes, as Thomson did, that a heat engine, operating reversibly and cyclically, produces work and transports *something* without diminution, here denoted Q, between its hot and cold heat reservoirs. Interestingly, although Thomson called this something *heat*, he did not require that this quantity have any of the properties of heat other than its conservation.

4.3 Carnot's Function Revealed

Recall that Carnot had proved that the duty or efficiency of a reversible, cyclic heat engine operating between two heat reservoirs depends only on the temperatures of the reservoirs. This efficiency, in terms of the empirical temperatures t_C and t_H of the two heat reservoirs, is expressed formally by

$$\frac{W}{Q} = \varepsilon(t_H, t_C) \tag{4.1}$$

where $\varepsilon(t_H, t_C)$ is called Carnot's function. Equation (4.1) reproduces equation (3.1).

According to Thomson's definition of absolute temperature, doubling the interval of absolute temperature between two heat reservoirs doubles the work produced by a Carnot engine operating between these two reservoirs. Furthermore, tripling this interval produces triple the work and so on. Therefore, according to Thomson's definition, the Carnot efficiency W/Q_H is, by design, directly proportional to the interval between the absolute temperatures of the reservoirs between which the engine operates. Thus, equation (4.1) may be expressed as

$$\frac{W}{Q} = C \cdot (T_H - T_C) \tag{4.2}$$

where C is a constant independent of the temperature interval $T_H - T_C$. Therefore, the relation between empirical and absolute temperatures is given by

$$\varepsilon(t_H, t_C) = C \cdot (T_H - T_C). \tag{4.3}$$

Thomson's definition at least partly reveals Carnot's function in terms of absolute temperatures.

4.4 Prelude and Postlude

John Dalton had tried to construct an absolute scale of temperatures even earlier than Thomson did. Dalton's scale was based on Newton's law of cooling.[5] Yet Dalton failed, while Thomson

5. Cardwell, *From Watt to Clausius,* 125–126. Newton himself used "Newton's law of cooling" to denominate temperatures. See his report, "A Scale of the Degrees of Heat," in Magie, *Source Book in Physics*, 125–128.

succeeded. After all, Newton's law of cooling is only an approximation that works more or less well depending on the properties of the materials between which heat is conducted, while the operation of a Carnot engine is an ideal, independent of all material properties.

The consequences of Thomson's 1848 definition of absolute temperature may never have been fully explored. For almost immediately, upon its proposal, Thomson became troubled by James Joule's experimental demonstrations that work is converted into heat at a fixed rate—an idea that meant that the heat flowing *into* a Carnot engine may not equal the heat flowing *out* of it. And if these two quantities are not equal, Thomson would have to rethink Carnot's heat engine analysis and reformulate his definition of absolute temperature.

Thomson's paper is here reproduced in full. The paragraph giving an operational definition of his absolute scale begins, about halfway through the paper, with the phrase, "The characteristic property of the scale which I now propose is, that all degrees have the same value."—DSL

William Thomson (Kelvin), 1848

"On an Absolute Thermometric Scale Founded on Carnot's Theory of the Motive Power of Heat,* and Calculated from

*Published in 1824 in a work entitled *Reflexions sur la Puissance Motrice du Feu,* by M. S. Carnot. Having never met with the original work, it is only through a paper by M. Clapeyron, on the same subject, published in the *Journal de Ecole Polytechnique,* Vol. xiv. 1834, and translated in the first volume of *Taylor's Scientific Memoir,* that the Author has become acquainted

Regnault's Observation,"[†] *Mathematical and Physical Papers*, vol. 1, article 39 (Cambridge: Cambridge University Press, 1882), 100–106.

The determination of temperature has long been recognized as a problem of the greatest importance in physical science. It has accordingly been made a subject of most careful attention, and, especially in late years, of very elaborate and refined experimental researches;[‡] and we are thus at present in possession of as complete a practical solution of the problem as can be desired, even for the most accurate investigations. The theory of thermometry is however as yet far from being in so satisfactory a state. The principle to be followed in constructing a thermometric scale might at first sight seem to be obvious, as it might appear that a perfect thermometer would indicate equal additions of heat, as corresponding to equal elevations of temperature, estimated by the numbered divisions of its scale. It is however now recognized (from the variations in the specific heats of bodies) as an experimentally demonstrated fact that thermometry under this condition is impossible, and we are left without any principle on which to found an absolute thermometric scale.

---

with Carnot's Theory.—W. T. [Note of Nov. 5th, 1851. A few months later through the kindness of my late colleague Prof. Lewis Gordon, I received a copy of Carnot's original work and was thus enabled to give to the Royal Society of Edinburgh my "Account of Carnot's theory" which is reprinted as Art. XLI below. The original work has since been republished, with a biographical notice, Paris, 1878.]

[†] An account of the first part of a series of researches undertaken by M. Regnault by order of the French Government, for ascertaining the various physical data of importance in the Theory of the Steam Engine, is just published in the *Memoires d'Institut*, of which it constitutes the twenty-first volume (1847). The second part of the researches has not yet been published. [Note of Nov. 5, 1881. The continuation of these researches has now been published: thus we have for the whole series, Vol. I. in 1847; Vol. II. in 1862; and Vol. III. in 1870.]

[‡] A very important section of Regnault's work is devoted to this object.

Next in importance to the primary establishment of an absolute scale, independently of the properties of any particular kind of matter, is the fixing upon an arbitrary system of thermometry, according to which results of observations made by different experimenters, in various positions and circumstances, may be exactly compared. This object is very fully attained by means of thermometers constructed and graduated according to the clearly defined methods adopted by the best instrument-makers of the present day, when the rigorous experimental processes which have been indicated, especially by Regnault, for interpreting their indications in a comparable way, are followed. The particular kind of thermometer which is least liable to uncertain variations of any kind is that founded on the expansion of air, and this is therefore generally adopted as the standard for the comparison of thermometers of all constructions. Hence the scale which is at present employed for estimating temperature is that of the air thermometer; and in accurate researches care is always taken to reduce to this scale the indications of the instrument actually used, whatever may be its specific construction and graduation.

The principle according to which the scale of the air thermometer is graduated, is simply that equal absolute expansions of the mass of air or gas in the instrument, under a constant pressure, shall indicate equal differences of the numbers on the scale; the length of a "degree" being determined by allowing a given number for the interval between the freezing- and the boiling-points. Now it is found by Regnault that various thermometers, constructed with air under different pressures, or with different gases, give indications which coincide so closely, that, unless when certain gases, such as sulphurous acid, which approach the physical condition of vapours at saturation, are made use of, the variations are inappreciable.* This remarkable circumstance enhances very

*Regnault, *Relation des Experiences*, &c., Fourth Memoir, First Part. The differences, it is remarked by Regnault, would be much more sensible if the graduation were effected on the supposition that the coefficients of

much the practical value of the air-thermometer; but still a rigorous standard can only be defined by fixing upon a certain gas at a determinate pressure, as the thermometric substance. Although we have thus a strict principle for constructing a definite system for the estimation of temperature, yet as reference is essentially made to a specific body as the standard thermometric substance, we cannot consider that we have arrived at an absolute scale, and we can only regard, in strictness, the scale actually adopted as *an arbitrary series of numbered points of reference sufficiently close for the requirements of practical thermometry.*

In the present state of physical science, therefore, a question of extreme interest arises: Is there any principle on which an absolute thermometric scale can be founded? It appears to me that Carnot's theory of the motive power of heat enables us to give an affirmative answer.

The relation between motive power and heat, as established by Carnot, is such that *quantities of heat, and intervals of temperature* are involved as the sole elements in the expression for the amount of mechanical effect to be obtained through the agency of heat; and since we have, independently, a definite system for the measurement of quantities of heat, we are thus furnished with a measure for intervals according to which absolute differences of temperature may be estimated. To make this intelligible, a few words in explanation of Carnot's theory must be given; but for a full account of this most valuable contribution to physical science, the reader is referred to either of the works mentioned above (the original treatise by Carnot, and Clapeyron's paper on the same subject).

In the present state of science no operation is known by which heat can be absorbed, without either elevating the temperature of matter, or becoming latent and producing some alteration in the physical condition of the body into which it is absorbed; and the

---

expansion of the different gases are equal, instead of being founded on the principle laid down in the text, according to which the freezing- and boiling-points are experimentally determined for each thermometer.

conversion of heat (or caloric) into mechanical effect is probably impossible,[†] certainly undiscovered. In actual engines for obtaining mechanical effect through the agency of heat, we must consequently look for the source of power, not in any absorption and conversion, but merely in a transmission of heat. Now Carnot, starting from universally acknowledged physical principles, demonstrates that it is by the letting down of heat from a hot body to a cold body, through the medium of an engine (a steam engine, or an air engine for instance), that mechanical effect is to be obtained; and conversely, he proves that the same amount of heat may, by the expenditure of an equal amount of laboring force, be raised from the cold to the hot body (the engine being in this case worked backwards); just as mechanical effect may be obtained by the descent of water let down by a water-wheel, and by spending laboring force in turning the wheel backwards, or in working a pump, water may be elevated to a higher level. The amount of mechanical effect to be obtained by the transmission of a given quantity of heat, through the medium of any kind of engine in which the economy is perfect, will depend, as Carnot demonstrates, not on the specific nature of the substance employed as the medium of transmission of heat in the engine, but solely on the interval between the temperature of the two bodies between which the heat is transferred.

Carnot examines in detail the ideal construction of an air engine and of a steam engine, in which, besides the condition of perfect economy being satisfied, the machine is so arranged, that at the close of a complete operation the substance (air in one case and water in the other) employed is restored to precisely the same

[†]This opinion seems to be nearly universally held by those who have written on the subject. A contrary opinion however has been advocated by Mr. Joule of Manchester; some very remarkable discoveries which he has made with reference to the generation of heat by the friction of fluids in motion, and some known experiments with magneto-electric machines, seeming to indicate an actual conversion of mechanical effect into caloric. No experiment however is adduced in which the converse operation is exhibited; but it must be confessed that as yet much is involved in mystery with reference to these fundamental questions of natural philosophy.

physical condition as at the commencement. He thus shows on what elements, capable of experimental determination, either with reference to air, or with reference to a liquid and its vapor, the absolute amount of mechanical effect due to the transmission of a unit of heat from a hot body to a cold body, through any given interval of the thermometric scale, may be ascertained. In M. Clapeyron's paper various experimental data, confessedly very imperfect, are brought forward, and the amounts of mechanical effect due to a unit of heat descending a degree of the air- thermometer, in various parts of the scale, are calculated from them, according to Carnot's expressions. The results so obtained indicate very decidedly, that what we may with much propriety call the value of a degree (estimated by the mechanical effect to be obtained from the descent of a unit of heat through it) of the air thermometer depends on the part of the scale in which it is taken, being less for high than for low temperatures.[‡]

The characteristic property of the scale which I now propose is, that all degrees have the same value; that is, that a unit of heat descending from a body A at the temperature $T°$ of this scale, to a body B at the temperature $(T-1)°$, would give out the same mechanical effect, whatever be the number T. This may justly be termed an absolute scale, since its characteristic is quite independent of the physical properties of any specific substance.

To compare this scale with that of the air thermometer, the values (according to the principle of estimation stated above) of degrees of the air thermometer must be known. Now an expression, obtained by Carnot from the consideration of his ideal steam engine, enables us to calculate these values, when the

[‡]This is what we might anticipate, when we reflect that infinite cold must correspond to a finite number of degrees of the air-thermometer below zero; since, if we push the strict principle of graduation, stated above, sufficiently far, we should arrive at a point corresponding to the volume of air being reduced to nothing, which would be marked as –273° of the scale (–100/ · 366, if · 366 be the coefficient of expansion); and therefore –273° of the air thermometer is a point which cannot be reached at any finite temperature, however low.

latent heat of a given volume and the pressure of saturated vapor at any temperature are experimentally determined. The determination of these elements is the principal object of Regnault's great work, already referred to, but at present his researches are not complete. In the first part, which alone has been as yet published, the latent heats of a given weight, and the pressures of saturated vapor, at all temperatures between 0° and 230° (Cent. of the air thermometer), have been ascertained; but it would be necessary in addition to know the densities of saturated vapor at different temperatures, to enable us to determine the latent heat of a given volume at any temperature. M. Regnault announces his intention of instituting researches for this object; but till the results are made known, we have no way of completing the data necessary for the present problem, except by estimating the density of saturated vapor at any temperature (the corresponding pressure being known by Regnault's researches already published) according to the approximate laws of compressibility and expansion (the laws of Mariotte and Gay-Lussac, or Boyle and Dalton). Within the limits of natural temperature in ordinary climates, the density of saturated vapor is actually found by Regnault (Etudes Hygrométriques in the *Annales de Chimie*) to verify very closely these laws; and we have reason to believe from experiments which have been made by Gay-Lussac and others, that as high as the temperature 100° there can be no considerable deviation; but our estimate of the density of saturated vapor, founded on these laws, may be very erroneous at such high temperatures as 230°. Hence a completely satisfactory calculation of the proposed scale cannot be made till after the additional experimental data shall have been obtained; but with the data which we actually possess, we may make an approximate comparison of the new scale with that of the air-thermometer, which at least between 0° and 100° will be tolerably satisfactory.

The labor of performing the necessary calculations for effecting a comparison of the proposed scale with that of the air thermometer, between the limits 0° and 230° of the latter, has been kindly undertaken by Mr. William Steele, lately of Glasgow College, now of St Peter's College, Cambridge. His results in tabulated forms were laid before the Society, with a diagram, in which the comparison between the two scales is represented graphically. In the first table, the amounts of mechanical effect due to the descent of a unit of heat through the successive degrees of the air thermometer are exhibited. The unit of heat adopted is the quantity necessary to elevate the temperature of a kilogram of water from 0° to 1° of the air thermometer; and the unit of mechanical effect is a meter-kilogram; that is, a kilogram raised a meter high.

In the second table, the temperatures according to the proposed scale, which correspond to the different degrees of the air thermometer from 0° to 230°, are exhibited. [The arbitrary points which coincide on the two scales are 0° and 100°.]

Note—If we add together the first hundred numbers given in the first table, we find 135.7 for the amount of work due to a unit of heat descending from a body A at 100° to B at 0°. Now 79 such units of heat would, according to Dr. Black (his result being very slightly corrected by Regnault), melt a kilogram of ice. Hence if the heat necessary to melt a pound of ice be now taken as unity, and if a meter pound be taken as the unit of mechanical effect, the amount of work to be obtained by the descent of a unit of heat from 100° to 0° is 79 x 135.7, or 10,700 nearly. This is the same as 35,100 foot pounds, which is a little more than the work of a one horse-power engine (33,000 foot pounds) in a minute; and consequently, if we had a steam-engine working with perfect economy at one horse-power, the boiler being at the temperature 100°, and the condenser kept at 0° by a constant supply of ice, rather less, than a pound of ice would be melted in a minute.

[Note of Nov. 4, 1881. This paper was wholly founded on Carnot's uncorrected theory, according to which the quantity of heat taken in in the hot part of the engine, (the boiler of the steam engine for instance), was supposed to be equal to that abstracted from the cold part (the condenser of the steam engine), in a complete period of the regular action of the engine, when every varying temperature, in every part of the apparatus, has become strictly periodic. The reconciliation of Carnot's theory with what is now known to be the true nature of heat is fully discussed in Article XLVIII below; and in §§ 24–41 of that article, are shown in detail the consequently required corrections of the thermodynamic estimates of the present article. These corrections however do not in any way affect the absolute scale for thermometry which forms the subject of the present article. Its relation to the practically more convenient scale (agreeing with air thermometers nearly enough for most purposes, throughout the range from the lowest temperatures hitherto measured, to the highest that can exist so far as we know) which I gave subsequently. Dynamical Theory of Heat (Art. XLVIII. below), I'art vi., g§ 99, 100; Trans. R. S. E., May, 1854: and Article—"Heat," §§ 35–38, 47–67, Encyclopedia Britannica is shown in the following formula:

$$\theta = 100 \frac{\log t - \log 273}{\log 373 - \log 273}$$

where θ and t are the reckonings of one and the same temperature, according to my first and according to my second thermodynamic absolute scale.]

5
Mechanical Equivalent of Heat

5.1 Caloric: Conserved or Consumed?

Carnot held that all the caloric entering a heat engine from the hotter reservoir had to be exhausted to the colder reservoir before the working fluid of the engine could return to its initial condition. Thus, he wrote, "The production of motive power is ... due in steam engines not to an actual consumption of caloric but to its transmission from a warm body to a cold body."[1] According to Carnot this understanding "has never been called into question. ... To deny it would be to overthrow the whole theory of heat to which it serves as a basis."[2] Even so, the proposition that caloric is conserved as it passes through a heat engine seemed to play no role in the logic of Carnot's analysis—as recapitulated in chapter 3.

Another of Carnot's statements, less frequently remarked on, does play an important role. Recall Carnot's theorem, according to which no heat engine operating in a cycle between two heat

1. Sadi Carnot, *Reflections on the Motive Power of Fire and Other Papers on the Second Law of Thermodynamics* (Gloucester, MA: Peter Smith, 1977, 1960), 7.
2. Ibid., 19.

reservoirs can be more efficient than a reversible one, for a yet more efficient heat engine would create

> ... not only perpetual motion, but an unlimited creation of motive power without consumption either of caloric or of any other agent whatever. Such a creation is entirely contrary to ideas now accepted, to the laws of mechanics and of sound physics. It is inadmissible.[3]

Carnot's characterization of a heat engine that creates motive power without consuming caloric (or of any other agent) as *inadmissible* is the crucial step in his indirect proof of the theorem now known as *Carnot's theorem*.

Interestingly, in these two passages from the *Reflections*, the word *consumption* is used in apparently contradictory ways. In the first, a heat engine does not consume caloric, while in the second, caloric (or some other agent) must be consumed. Presumably the first reference is to caloric as a material substance, while the second is to its relative position in the hotter reservoir but not in the colder one. The first kind of caloric is conserved, while the second is consumed. At least, this is one way of harmonizing these two passages.

What we do know is that soon after composing the *Reflections*, Carnot began to consider that in a heat engine, caloric indeed might be consumed and converted into work. Carnot's words in these two passages and the structure of his argument in the *Reflections* foreshadow this new idea. But when Carnot published the *Reflections* in 1824, the conversion of heat into work was more hypothesis than theory. What Carnot lacked was a self-consistent, quantifiable theory that could replace the relatively successful doctrine of the conservation of caloric.

3. Carnot, *Reflections*, 12.

5.2 Julius Robert von Mayer

Julius Robert von Mayer (1814–1878), a young physician from Heilbronn (in modern Germany), provided that theory. While on a hundred-day voyage to the Dutch East Indies, Mayer, who was the ship's doctor, noticed, while bleeding sailors under his charge, the "lighter-than-expected color of the venous blood of Europeans recently arrived in the tropics."[4] Evidently oxidation darkens the blood as a consequence of producing heat and the body needs less heat in warm climes than in cold ones. For this reason, the venous blood of his sailors was less dark (or "lighter-than-expected") in the tropics than it was in Europe. This observation and its interpretation intrigued Mayer and sparked his interest in heat and its transformations.

Back in Heilbronn, Mayer continued to study heat until he concluded that work is converted into heat in determinate amounts and, conversely, heat into work. He also concluded that both heat and work are related to something else that, when animating living organisms and physical systems, was (to employ Mayer's words) *convertible, indestructible,* and *imponderable.* When Mayer published these ideas in the *Annalen der Chemie and Pharmacie* in 1842,[5] he called this "something else" *force,* as did others of his era, an unfortunate use since *force* also refers, via Newton's second law, to that which always attends acceleration. But clearly Mayer used the word *force* in his 1842 paper to refer to what we now call *energy.*

4. Kenneth Caneva, *Robert Mayer and the Conservation of Energy* (Princeton: Princeton University Press, 1993), 7.

5. J. R. Mayer, "Remarks on the Forces of Inorganic Nature," *Annalen der Chemie and Pharmacie* 42 (May 1842): 233. For an English translation, see *London, Edinburgh, and Dublin Philosophical Magazine and Journal of Science,* series 4, vol. 24 (July–Dec. 1862): 371–377, or W. F. Magie, *A Source Book in Physics* (Cambridge, MA: Harvard University Press, 1963, 1935), 196–203.

According to Mayer, energy is both *indestructible* and *imponderable*. Because energy is *indestructible*, it is *conserved*. Yet the only examples of conserved quantities known to Mayer were those of material substances. But energy is also *imponderable*, literally something without weight or mass. Combining, as it did, these seemingly contradictory qualities, Mayer's *force* (our *energy*) was, in 1842, something new under the sun—something weird.

5.3 Mechanical Equivalent of Heat

Mayer concluded his 1842 paper with a sketch of a derivation of "the quantity of heat which corresponds to a given quantity of motion or falling force." The ratio of these two quantities is what today we call the *mechanical equivalent of heat*. Mayer filled out his sketch three years later (in 1845) in a self-published pamphlet. These dates and circumstances became important in the priority dispute that developed between the partisans of Mayer and those of his near contemporary, James Joule (1818–1889).

Gases have two heat capacities: a heat capacity at constant volume C_V and a somewhat larger heat capacity at constant pressure C_P. Mayer reasoned that the latter is larger than the former because, if allowed, a heated gas expands by pushing against the constant pressure of the atmosphere. In this way, heat may not only raise the temperature of a gas; some of the heat is also converted into the work done in expanding the volume of the gas. Therefore, the difference between these two heat capacities should be a measure of the mechanical equivalent of heat.

The heat capacity at constant volume C_V is the ratio of the heat Q added to a gas having constant volume V to its consequent increase in temperature Δt. Thus, $C_V = Q/\Delta t$, as diagrammed in figure 5.1a. According to Mayer, extra heat δQ is needed

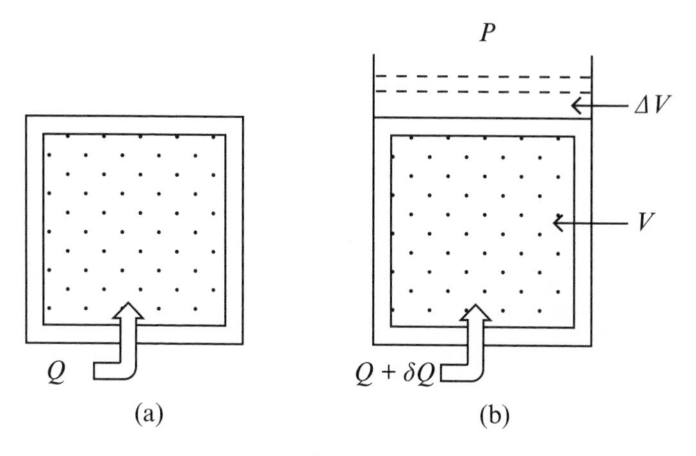

Figure 5.1
(a) $C_V = Q/\Delta t$. (b) $C_P = (Q + \delta Q)/\Delta t$. In each case, the same temperature increment Δt is realized. The container in panel (a) is rigid, while in panel (b) its top wall changes position in order to equalize the pressure P on the container.

to achieve the same temperature increase Δt when the gas is allowed to expand against a constant pressure P. Therefore, the heat capacity at constant pressure is given by $C_P = (Q + \delta Q)/\Delta t$, as illustrated in figure 5.1b.

The extra heat δQ is converted into the work $P\Delta V$ that expands the gas volume V by ΔV against the constant pressure P. Therefore, the ratio sought, that is, the mechanical equivalent of heat, usually symbolized with J, is defined by $J = P\Delta V/\delta Q$.[6]

6. The symbol J for the mechanical equivalent of heat anachronistically references James Joule (1818–1889), who indeed devoted his life to determining ever more accurate values of this ratio. Nevertheless, Mayer was the first to identify the concept of the mechanical equivalent of heat and determine a value for it.

The difference in the two heat capacities, $C_P - C_V = \delta Q/\Delta t$, produces an expression for the extra heat $\delta Q[= \Delta t(C_P - C_V)]$ required. Therefore,

$$J = \frac{P\Delta V}{\Delta t(C_P - C_V)}. \tag{5.1}$$

Assuming that the gas is ideal, and so observes the equation of state $PV = nR(273 + t)$ simplifies expression (5.1). Here t is the gas temperature in degrees centigrade, n is the number of moles of the gas, and R is the *ideal gas constant*. Thus, given the ideal gas law, we have $P\Delta V = nR\Delta t$ and equation (5.1) becomes[7]

$$J = \frac{nR}{(C_P - C_V)}. \tag{5.2}$$

Therefore, one needs only numerical values of the molar-specific heats, C_P/n and C_V/n, and the ideal gas constant R in order to determine the value of the mechanical equivalent of heat J. The data available to Mayer implied a value of J, a little less than that James Joule determined more directly and more accurately a few years later.

Mayer's derivation of equation (5.2) assumes that no heat is stored latently in the gas as it expands. Of course, latent heat would be stored if, as those who held to the doctrine of caloric believed, the particles of caloric and the particles of gas attract one another. To justify his assumption, Mayer appealed to an 1807 experiment of Joseph Louis Gay-Lussac in which the mere expansion of a gas into a vacuum caused no change in its

7. Today it is routine to assume that heat and work are denominated with the same unit. This is equivalent to setting J in equation (5.2) equal to the numeral 1. In this case equation (5.2) reduces to $C_P - C_V = nR$.

temperature.[8] Mayer's interpretation of Gay-Lussac's experiment was that no heat is stored latently in an expanding gas.

Mayer's arguments were sound. Yet he had no formal training in physics, had no academic connections or financial means, and could do no convincing experiments. (On one occasion, Mayer did find that the temperature of water "when violently shaken" increased from 12° to 13° C.)[9] And since he had no opportunity to publish except, at first, obscurely and then later privately, Mayer worked outside of and unknown to the small circle of natural philosophers concerned with heat.

Even when James Joule and William Thomson become aware of Mayer's work, they misunderstood his derivation and faulted him for making no direct measurements. Mayer, however, recognized Joule's independent discovery and measurement of the mechanical equivalent of heat but insisted on the priority of his own contribution. Unfortunately, Mayer's claim was ignored or criticized by those who should have known better and ridiculed by others who held dogmatically to the doctrine of caloric. Mayer, who longed for recognition, became depressed and in May 1850 threw himself from a third-floor window. He survived the fall and was involuntarily committed to a mental institution for a time.

Mayer recovered his health and continued to practice medicine but never recovered his zeal for science.[10] Fortunately, he

8. Morton Mott-Smith, *The Concept of Energy, Simply Explained* (New York: Dover, 1964), 85, and D. S. L. Cardwell, *From Watt to Clausius: The Rise of Thermodynamics in the Early Industrial Age* (Ithaca, NY: Cornell University Press, 1971), 234. Gay-Lussac's own account is found in Magie, *Source Book in Physics*, 170–172. Decades later, James Joule, who was unaware of Gay-Lussac's earlier experiment, also studied the free expansion of a gas. Today such free expansion is called *Joule expansion*.
9. Magie, *Source Book in Physics*, 201.
10. Mott-Smith, *Concept of Energy*, 97, and Kenneth L. Caneva, *Robert Mayer and the Conservation of Energy* (Princeton: Princeton University Press, 1993), 9.

lived long enough to see his work become widely known, fairly evaluated, and justly appreciated. He received Britain's highest award for scientific achievement, the Copley Medal of the Royal Society of London, in 1871, one year after Joule was similarly recognized. Mayer died in 1878.

5.4 James Joule

James Joule (1818–1889), like Mayer, was an academic outsider who struggled for recognition. However, Joule inherited and continued to manage his father's successful brewing business and so had the advantages that wealth can buy: a good education with private tutors, a well-furnished laboratory, and the leisure to perform experiments. And, perhaps not unrelated to his wealth, Joule was not easily discouraged.

Joule's interest in heat originated in the hope, unrealized in his lifetime, that electrical generators would replace the steam engines that powered the machinery in England's workshops. In the course of his investigations Joule made, in 1841, a significant discovery: the electrical current in a circuit produces heat distributed throughout the circuit at a rate equal to the product of the resistance of the circuit and the square of its current, an effect now called *Joule heating*. By 1843, he was convinced that this heat was created by the "chemical reactions in the battery" or, when the current was mechanically generated, by the "mechanical power exerted in turning a magneto-electric machine."[11]

11. J. P. Joule, "On the Caloric Effects of Magneto-Electricity, and on the Mechanical Value of Heat," *Philosophical Magazine* 23 (1843): 263, 347, 435, *Scientific Papers*, 123, as quoted in Cardwell, *From Watt to Clausius*, 232–233.

Joule measured the rate at which this work was converted into heat in terms of the peculiar British units of his time: the work required to lift a 1 pound weight 1 foot (a *foot-pound*) and the heat required to increase the temperature of 1 pound of water 1 degree Fahrenheit (a *British thermal unit, that is, a BTU*). His 1843 measurement of the mechanical equivalent of heat, 838 foot-pounds of work per BTU,[12] is somewhat higher than the value inscribed on his gravestone as summarizing his life's work: 772.55.[13]

Joule reported this measurement at a meeting of the British Association for the Advancement of Science. Although it attracted no attention, Joule persevered. In his own words,

> I shall lose no time in repeating and extending these experiments, being satisfied that the grand agents of nature are, by the Creator's fiat, *indestructible*; and that wherever mechanical force is expended, an exact equivalent of heat is always obtained. (1843)[14]

Yet if "wherever mechanical force is expended, an exact equivalent of heat is always obtained," heat is not conserved, and so, perhaps, it is also consumed whenever an equivalent of work is produced. Thus, not only did Joule, like Mayer, believe that work is converted into heat at a fixed rate but also that heat is converted into work at this rate in a heat engine.

12. In today's units, 4.48 Joules/calorie.

13. In 1843 Ludwig A. Colding, a Danish engineer and physicist, also independently demonstrated the existence of a mechanical equivalent of heat by showing that in a given experiment, the work consumed and the heat produced were directly proportional to one another. See especially pages 8–9 of Kenneth Caneva, "Colding, Ørsted, and the Meanings of Force," *Historical Studies in the Physical and Biological Sciences* 28, no. 1 (1997): 1–138.

14. Joule, "On the Caloric Effects of Magneto-Electricity," 263–276, 347–355, 435–443, and as quoted in Magie, *Source Book in Physics*, 205.

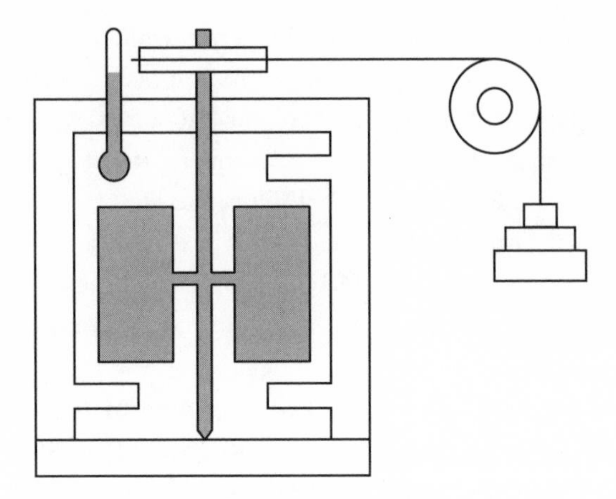

Figure 5.2
Joule's apparatus for measuring the mechanical equivalent of heat.

Joule's most dramatic and transparent apparatus for measuring the rate at which work is converted into heat allowed a falling weight to turn vanes that churned a liquid (water, sperm whale oil, or mercury) within an insulated container, as illustrated in figure 5.2. The total distance the weight, which was repeatedly restored to its initial position, fell determined the work done on the liquid, while the increase in the liquid temperature and its specific heat determined the heat generated. In 1845 Joule reported a value for the mechanical equivalent of heat, produced with this apparatus.[15] Joule's work was again ignored. But he continued to vary and repeat his experiments and make ever more accurate and precise measurements.

15. J. P. Joule, "On the Existence of an Equivalent Relation between Heat and the Ordinary Forms of Mechanical Power," *Philosophical Magazine* 27, 3rd series (1845): 205.

Then in 1847, at another meeting of the British Association, the young William Thomson rose to question Joule. A discussion ensued. Thomson was open to new ideas and grasped Joule's claim that heat and work were mutually convertible even as he, at the time, rejected the idea. Thomson was, after all, committed to Carnot's analysis of heat engine operation, and Carnot's analysis, at least superficially, depended on the conservation of caloric. But Joule's 1847 encounter with Thomson marked an important transition in the history of thermodynamics—to an unstable phase in which Carnot's analysis and the mutual convertibility of heat and work clashed. Neither could be denied. Yet the two contradicted each other—or so it seemed.

If the style of the following paper, larded as it is with vague principles such as the "indestructibility of causes," with Latin maxims and arcane concepts, and with the use of "force" for "energy," confuses modern readers, Mayer's contemporaries were similarly confused. Nevertheless, a careful study of this paper will reveal that Mayer grasped the idea of the conservation of energy. Its last two paragraphs refer to the outline of a calculation of the mechanical equivalent of heat that Mayer more fully explained three years later.—DSL

Robert von Mayer, 1842

"Remarks on the Forces of Inorganic Nature," *London, Edinburgh, and Dublin Philosophical Magazine and Journal of Science,* series 4, vol. 24 (July–Dec. 1862): 371–377. Translated from the German by G. C. Foster, B.A., Lecturer on Natural Philosophy.

The following pages are designed as an attempt to answer the questions, What are we to understand by "Forces" and how are different forces related to each other? Whereas the term *matter* implies the possession, by the object to which it is applied, of very definite properties, such as weight and extension; the term *force* conveys for the most part the idea of something unknown, unsearchable, and hypothetical. An attempt to render the notion of force equally exact with that of matter, and so to denote by it only objects of actual investigation, is one which, with the consequences that flow from it, ought not to be unwelcome to those who desire that their views of nature may be clear and unencumbered by hypotheses.

Forces are causes: accordingly, we may in relation to them make full application of the principle—*causa æquat effectum*. If the cause c has the effect e, then $c = e$; if, in its turn, e is the cause of a second effect f, we have $e = f$, and so on: $c = e = f \ldots = c$. In a chain of causes and effects, a term or a part of a term can never, as plainly appears from the nature of an equation, become equal to nothing. This first property of all causes we call their *indestructibility*.

If the given cause c has produced an effect e equal to itself, it has in that very act ceased to be: c has become e; if, after the production of e, c still remained in whole or in part, there must be still further effects corresponding to this remaining cause: the total effect of c would thus be $> e$, which would be contrary to the supposition $c = e$. Accordingly, since c becomes e, and e becomes f, &c., we must regard these various magnitudes as different forms under which one and the same object makes its appearance. This capability of assuming various forms is the second essential property of all causes. Taking both properties together, we may say, causes are (quantitatively) *indestructible* and (qualitatively) *convertible* objects.

Two classes of causes occur in nature, which, so far as experience goes, never pass one into another. The first class consists of such causes as possess the properties of weight and impenetrability; these are kinds of Matter: the other class is made up of causes which are wanting in the properties just mentioned, namely Forces, called also Imponderables, from the negative property that has been indicated. Forces are therefore *indestructible, convertible, imponderable* objects.

We will in the first instance take matter, to afford us an example of causes and effects. Explosive gas, H+0, and water, HO, are related to each other as cause and effect, therefore H+O=H0. But if H+O becomes HO, heat, *cal.,* makes its appearance as well as water; this heat must likewise have a cause, *x*, and we have therefore H+O+x=HO+*cal.* It might, however, be asked whether H+O is really =H0, and x=*cal.*, and not perhaps H+O=*cal.*, and x=H0, whence the above equation could equally be deduced; and so in many other cases. The phlogistic chemists recognized the equation between cal. and x, or Phlogiston as they called it, and in so doing made a great step in advance; but they involved themselves again in a system of mistakes by putting –x in place of O; thus, for instance, they obtained H = HO+x.

Chemistry, whose problem it is to set forth in equations the causal connection existing between the different kinds of matter, teaches us that matter, as a cause, has matter for its effect; but we are equally justified in saying that to force as cause, corresponds force as effect. Since $c = e$, and $e = c$, it is unnatural to call one term of an equation a force, and the other an effect of force or phenomenon, and to attach different notions to the expressions Force and Phenomenon. In brief, then, if the cause is matter, the effect is matter; if the cause is a force, the effect is also a force.

A cause which brings about the raising of a weight is a force; its effect (*the raised weight*) is, accordingly, equally a force; or, expressing this relation in a more general form, separation in space of ponderable objects is a force; since this force causes the fall of bodies, we call it falling force. Falling force and fall, or, more generally still, falling force and motion, are forces which are related to each other as cause and effect—forces which are convertible one into the other—two different forms of one and the same object. For example, a weight resting on the ground is not a force: it is neither the cause of motion, nor of the lifting of another weight; it becomes so, however, in proportion as it is raised above the ground: the cause—the distance between a weight and the earth—and the effect—the quantity of motion produced—bear to each other, as we learn from mechanics, a constant relation.

Gravity being regarded as the cause of the falling of bodies, a gravitating force is spoken of, and so the notions of property and of force are confounded with each other: precisely that which is the essential attribute of every force—the union of indestructibility with convertibility—is wanting in every property: between a property and a force, between gravity and motion, it is therefore impossible to establish the equation required for a rightly conceived causal relation. If gravity be called a force, a cause is supposed which produces effects without itself diminishing, and incorrect conceptions of the causal connection of things are thereby fostered. In order that a body may fall, it is no less necessary that it should be lifted up, than that it should be heavy or possess gravity; the fall of bodies ought not therefore to be ascribed to their gravity alone.

It is the problem of Mechanics to develop the equations which subsist between falling force and motion, motion and falling force, and between different motions: here we will call to mind only one point. The magnitude of the falling force v is directly proportional (the earth's radius being assumed $= \infty$) to the magnitude of the mass m, and the height d to which it is raised; that is, $v = md$. If the height $d = 1$, to which the mass m is raised, is transformed into the final velocity $c = 1$ of this mass, we have also $v = mc$, but from the known relations existing between d and c, it results that, for other values of d or of c, the measure of the force v is mc^2; accordingly $v = md = mc^2$: the law of the conservation of *vis viva* is thus found to be based on the general law of the indestructibility of causes.

In numberless cases we see motion cease without having caused another motion or the lifting of a weight; but a force once in existence cannot be annihilated, it can only change its form; and the question therefore arises, What other forms is force, which we have become acquainted with as falling force and motion, capable of assuming? Experience alone can lead us to a conclusion on this point. In order to experiment with advantage, we must select implements which, besides causing a real cessation of motion, are as little as possible altered by the objects to be examined. If, for example, we rub together two metal plates, we see motion disappear, and heat, on the other hand, make its appearance, and we have now only to ask whether motion is the cause of heat. In order to come to a decision on this point, we must discuss the question whether, in the numberless cases in which the expenditure of motion is accompanied by the appearance of heat, the motion has not some other effect than the production of heat, and the heat some other cause than the motion.

An attempt to ascertain the effects of ceasing motion has never yet been seriously made; without, therefore, wishing to exclude *a priori* the hypotheses which it may be possible to set up, we observe only that, as a rule, this effect cannot be supposed to be an alteration in the state of aggregation of the moved (that is, rubbing, &c.) bodies. If we assume that a certain quantity of motion v is expended in the conversion of a rubbing substance m into n, we must then have $m + v = n$ and $n = m + v$; and when n is reconverted into m, v must appear again in some form or other. By the friction of two metallic plates continued for a very long time, we can gradually cause the cessation of an immense quantity of movement; but would it ever occur to us to look for even the smallest trace of the force which has disappeared in the metallic dust that we could collect, and to try to regain it thence? We repeat, the motion cannot have been annihilated; and contrary, or positive and negative, motions cannot be regarded as $= 0$, any more than contrary motions can come out of nothing, or a weight can raise itself.

Without the recognition of a causal connection between motion and heat, it is just as difficult to explain the production of heat as it is to give any account of the motion that disappears. The heat cannot be derived from the diminution of the volume of the rubbing substances. It is well known that two pieces of ice may be melted by rubbing them together *in vacuo*; but let any one try to convert ice into water by pressure, however enormous. Water undergoes, as was found by the author, a rise of temperature when violently shaken. The water so heated (from 12° to 13° C.) has a greater bulk after being shaken than it had before; whence now comes this quantity of heat, which by repeated shaking may be called into existence in the same apparatus as often as we please? The vibratory hypothesis of heat is an approach towards the doctrine of heat being the effect of motion, but it does not favor the admission of this causal relation in its full generality; it rather lays the chief stress on uneasy oscillations (*unbehagliche Schwingungen*).

If it be now considered as established that in many cases (*exceptio confirmat regulam*) no other effect of motion can be traced except heat, and that no other cause than motion can be found for the heat that is produced, we prefer the assumption that heat proceeds from motion, to the assumption of a cause without effect and of an effect without a cause—just as the chemist, instead of allowing oxygen and hydrogen to disappear without further investigation, and water to be produced in some inexplicable manner, establishes a connection between oxygen and hydrogen on the one hand and water on the other.

The natural connection existing between falling force, motion, and heat may be conceived of as follows. We know that heat makes its appearance when the separate particles of a body approach nearer to each other: condensation produces heat. And what applies to the smallest particles of matter, and the smallest intervals between them, must also apply to large masses and to measureable distances. The falling of a weight is a real diminution of the bulk of the earth, and must therefore without doubt be related to the quantity of heat thereby developed; this quantity of heat must be proportional to the greatness of the weight and its distance from the ground. From this point of view we are very easily led to the equations between falling force, motion, and heat that have already been discussed.

But just as little as the connection between falling force and motion authorizes the conclusion that the essence of falling force is motion, can such a conclusion be adopted in the case of heat. We are, on the contrary, rather inclined to infer that, before it can become heat, motion—whether simple, or vibratory as in the case of light and radiant heat, &c.—must cease to exist as motion.

If falling force and motion are equivalent to heat, heat must also naturally be equivalent to motion and falling force. Just as heat appears as an effect of the diminution of bulk and of the cessation of motion, so also does heat disappear as a cause when its effects are produced in the shape of motion, expansion, or raising of weight.

In water-mills, the continual diminution in bulk which the earth undergoes, owing to the fall of the water, gives rise to motion, which afterwards disappears again, calling forth unceasingly a great quantity of heat; and inversely, the steam-engine serves to decompose heat again into motion or the raising of weights. A locomotive engine with its train may be compared to a distilling apparatus; the heat applied under the boiler passes off as motion, and this is deposited again as heat at the axles of the wheels.

We will close our disquisition, the propositions of which have resulted as necessary consequences from the principle *causa sequat effectum,* and which are in accordance with all the phenomena of Nature, with a practical deduction. The solution of the equations subsisting between falling force and motion requires that the space fallen through in a given time, *e.g.* the first second, should be experimentally determined; in like manner, the solution of the equations subsisting between falling force and motion on the one hand and heat on the other, requires an answer to the question, How great is the quantity of heat which corresponds to a given quantity of motion or falling force? For instance, we must ascertain how high a given weight requires to be raised above the ground in order that its falling force may be equivalent to the raising of the temperature of an equal weight of water from 0° to 1° C. The attempt to show that such an equation is the expression of a physical truth may be regarded as the substance of the foregoing remarks.

By applying the principles that have been set forth to the relations subsisting between the temperature and the volume of gases, we find that the sinking of a mercury column by which a gas is compressed is equivalent to the quantity of heat set free by the compression; and hence it follows, the ratio between the capacity for heat of air under constant pressure and its capacity under constant volume being taken as =1.421, that the warming of a given weight of water from 0° to 1° C. corresponds to the fall of an equal weight from the height of about 365 meters. If we compare with this result the working of our best steam-engines, we see how small a part only of the heat applied under the boiler is really transformed into motion or the raising of weights; and this may serve as justification for the attempts at the profitable production of motion by some other method than the expenditure of the chemical difference between carbon and oxygen—more particularly by the transformation into motion of electricity obtained by chemical means.

Here Joule briefly describes his simplest and most transparent measurement of the mechanical equivalent of heat. Because he was an accomplished brewer, Joule was no doubt capable of measuring the small changes in temperature he described.—DSL

James Joule, 1845

"Letter to the Editor: On the Existence of an Equivalent Relation between Heat and the ordinary Forms of Mechanical Power," *Philosophical Magazine*, 3rd ser., 27 (1845): 205.

Gentlemen,

The principal part of this letter was brought under the notice of the British Association at its last meeting at Cambridge. I have hitherto hesitated to give it further publication, not because I was in any degree doubtful of the conclusions at which I had arrived, but because I intended to make a slight alteration in the apparatus calculated to give still greater precision to the experiments. Being unable, however, just at present to spare time necessary to fulfill this design, and being at the same time most anxious to convince the scientific world of the truth of the positions I have maintained, I hope you will do me the favor of publishing this letter in your excellent Magazine.

The apparatus exhibited before the Association consisted of a brass paddle-wheel working *horizontally* in a can of water. Motion could be communicated to this paddle by means of weights, pulleys, &c., exactly in the matter described in a previous paper.*

The paddle moved with great resistance in the can of water, so that the weights (each of four pounds) descended at the slow rate of about one foot per second. The height of the pulleys from the ground was twelve yards, and consequently, when the weights had descended through that distance, they had to be wound up again in order to renew the motion of the paddle. After this operation had been repeated sixteen times, the increase of the temperature of the water was ascertained by means of a very sensible and accurate thermometer.

A series of nine experiments was performed in the above manner, and nine experiments were made in order to eliminate the cooling or heating effects of the atmosphere. After reducing the result to the capacity for heat of a pound of water, it appeared that for each degree of heat evolved by the friction of water a mechanical power equal to that which can raise a weight of 890 lb to the height of one foot had been expended.

The equivalents I have already obtained are;—1st, 823 lb., derived from magneto-electrical experiments (Phil. Mag. ser. 3 vol. xxiii. pp. 263, 347); 2nd, 795 lb., deduced from the cold produced

*Phil. Mag. ser. 3, vol. xxiii, 436. The paddle-wheel used by Rennie in his experiments on the friction of water (Phil. Trans. 1831, plate xi, fig, 1) was somewhat similar to mine. I have employed, however, a greater number of "floats," and also a corresponding number of stationary floats, in order to prevent the rotatory motion of the can.

by the rarefaction of air (Ibid. May 1845, p. 369); and 3rd, 774 lb. from experiments (hitherto unpublished) on the motion of water through narrow tubes. This last class of experiments being similar to that with the paddle wheel, we may take the mean of 774 and 890, or 832 lb., as the equivalent derived from the friction of water. In such delicate experiments, where one hardly ever collects more than one another than that above exhibited could hardly have been expected. I may therefore conclude that the existence of an equivalent relation between heat and the ordinary forms of mechanical power is proved; and assume 817 lb., the mean of the results of three distinct classes of experiments, as the equivalent, until more accurate experiments shall have been made.

Any of your readers who are so fortunate as to reside amid the romantic scenery of Wales or Scotland could, I doubt not, confirm my experiments by trying the temperature of the water at the top and at the bottom of a cascade. If my views be correct, a fall of 817 feet will course generate one degree of heat, and the temperature of the river Niagra will be raised about one fifth of a degree by its fall of 160 feet.

Admitting the correctness of the equivalent I have named, it is obvious that the *vis viva* of the particles of a pound water at (say) 51° is equal to the *vis viva* possessed by a pound of water at 50° plus the *vis viva* which would be acquired by a weight of 817 lb. after falling through the perpendicular height of one foot.

Assuming that the expansion of elastic fluids on the removal of pressure is owing to the centrifugal force of revolving atmospheres of electricity, we can easily estimate the absolute quantity of heat in matter. For in an elastic fluid the pressure will be proportional to the square of the velocity of the revolving atmosphere, and the *vis viva* of the atmospheres will also be proportional to the square of their velocity; consequently the pressure of elastic fluids at the temperatures 32° and 33° is 480: 481; consequently the zero of temperature must be 480° below the freezing-point of water.

We see then what an enormous quantity of *vis viva* exists in matter. A single pound of water at 60° must possess 480° + 28° =

508° of heat; in other words, it must possess a *vis viva* equal to that acquired by a weight of 415036 lb. after falling through the perpendicular height of one foot. The velocity with which the atmosphere of electricity must revolve in order to present this enormous amount of *vis viva* must of course be prodigious, and equal probably to the velocity of light in the planetary space, or to that of an electric discharge as determined by the experiments of Wheatstone.

I remain, Gentlemen,
Yours Respectfully,

James P. Joule

6
First Law of Thermodynamics

6.1 Rudolf Clausius

In 1850 Rudolf Clausius (1822–1888) was ready to begin his first university position. Although a little older than William Thomson (1824–1907), Clausius was essentially Thomson's student.[1] Clausius had been studying Thomson's papers, in particular Thomson's 1849 account of Carnot's *Reflections*,[2] which Thompson had at the time recently inspected. In 1850, Clausius knew of Carnot's analysis of heat engine operation only through Thomson.

Clausius's first paper established his reputation as a cofounder, along with Thomson, of the new science of *thermodynamics*, a word Thomson invented. "On the Motive Power of Heat, and on the Laws Which Can Be Deduced from It for the Theory of Heat,"[3] the first of nine papers Clausius wrote on the

1. D. S. L. Cardwell, *From Watt to Clausius: The Rise of Thermodynamics in the Early Industrial Age* (Ithaca, NY: Cornell University Press, 1971), 244.
2. William Thomson, "An Account of Carnot's Theory of the Motive Power of Heat; with Numerical Results Deduced from Regnault's Experiments on Steam," *Royal Society of Edinburgh* 16, no. 5 (1849): 541–574.
3. Rudolf Clausius, Pogendorff's *Annalen der Physik* 79 (1850): 368, 500. See also Rudolf Clausius, *Reflections on the Motive Power of Fire* (Gloucester, MA: Peter Smith, 1977), 109–152.

subject,[4] accomplished two tasks. First, it harmonized Carnot's analysis with the convertibility of heat and work, and, second, it identified two principles or laws on which the science of thermodynamics could be founded. I have separated these contemporaneous identifications into two chapters—this one on the first law of thermodynamics and the next on the second law.

Clausius saw no conflict between Carnot's requirement that heat must flow from a hotter body to a colder one in order to produce work and Mayer and Joule's claim that heat is consumed when a heat engine produces work:

> It is not at all necessary to discard Carnot's theory entirely, a step which we certainly would find hard to take, since it has to some extent been conspicuously verified by experience. A careful examination shows that the new method does not stand in contradiction to the essential principle of Carnot, but only to the subsidiary statement that *no heat is lost,* since in the production of work it may very well be the case that at the same time a certain quantity of heat is consumed and another quantity transferred from a hotter to a colder body, and both quantities of heat stand in a definite relation to the work that is done. This will appear more plainly in the sequel. (1850)[5]

Thus, Clausius reimagined the flow of heat in a heat engine from one in which caloric is conserved, as in figure 3.3, to one in which part of the heat is consumed and converted into its work equivalent, as in figure 6.1. (The mechanical equivalent of heat allows us to denominate heat and work in the same unit, as we do in figure 6.1 and in the balance of this book.)

4. Collected in Rudolph Clausius, *The Mechanical Theory of Heat* (London: John Van Voorst, 1867).
5. Clausius, *Reflections on the Motive Power of Fire,* 112.

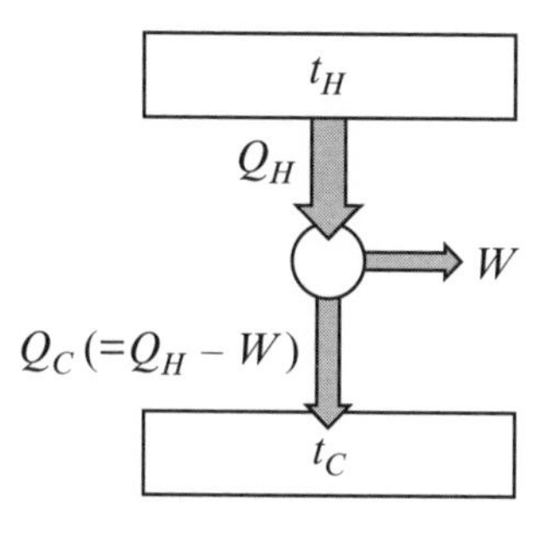

Figure 6.1
Heat flow in a heat engine as envisioned by Rudolf Clausius.

Clausius's "new method" replaces the conservation of caloric with what he called the "principle of the equivalence of heat and work," according to which:

> In all cases in which work is produced by the agency of heat, a quantity of heat is consumed which is proportional to the work done; and, conversely, by the expenditure of an equal quantity of work an equal quantity of heat is produced. (1850)[6]

6.2 System, State, and Boundary

Other important consequences of the principle of the equivalence of heat and work become apparent when applied to changes in the state of a finite system surrounded by special boundaries.

A *system* is that part of the universe with which one is concerned—say, a lump of iron, a bucket of water, or the air

6. Clausius, *Reflections*, 112.

within a cylindrical container. Systems are surrounded by *boundaries* that can prohibit or permit certain interactions. For instance, a rigid boundary prohibits mechanical work from being done on or by the system, a thermally insulating boundary prohibits heat transfer between system and environment, and an impermeable membrane prohibits the passage of matter.

A *thermodynamic system* is characterized by steady-state variables that include temperature. For instance, the *state variables* of an isotropic fluid are pressure, volume, temperature, and mole number. State variables are related to one another in *equations of state*, the best known of which, in 1850, were those of the ideal gas model.

Consider a thermodynamic system composed of the working fluid of a Carnot engine that returns to its initial state at the end of each cycle. This fluid is successively surrounded by boundaries that permit work interactions and alternately permit and prohibit heat interactions. During one cycle, the fluid receives heat Q_H from the hot heat reservoir, rejects heat Q_C to the cold reservoir, and performs work W on its environment, as illustrated in figure 6.1. According to Clausius's principle of the equivalence of heat and work, the heat rejected to the cold reservoir Q_C is equal to the heat received from the hot reservoir Q_H minus the work done by the fluid W during a complete cycle. Thus, $Q_C = Q_H - W$ or, equivalently,

$$0 = (Q_H - Q_C) - W \tag{6.1}$$

where $(Q_H - Q_C)$ represents the net heat *received by* the fluid from the environment and W is the net work the fluid *does on* the environment. The two terms on the right-hand side of equation (6.1) sum to zero only because they describe interactions in one cycle.

6.3 The First Law

The transformations of the working fluid in a heat engine during one cycle are a special case. During an arbitrary period, the signed quantity Q representing the net heat *received from* the environment and the signed quantity W representing the net work *done by* the working fluid (more briefly and more generally "the system") on its environment do not sum to zero. Thus, if $Q > W$, net "work-heat equivalence" is added to the system, and if $Q < W$, net work-heat equivalence is subtracted from the system. In both cases, the state of the system is changed.[7]

Because changing the work-heat equivalence of a system changes its state variables, a state variable with the dimension of Q and W, which we temporarily denote X, must exist whose net change,

$$\Delta X = Q - W, \tag{6.2}$$

is related through equations of state to changes in other state variables, for instance, to changes in the pressure, volume, and temperature. Therefore, the only purpose of ΔX is to quantify a system's change of state under work and heat interactions. Equation (6.2) follows directly from the equivalence of work and heat and that either may change a system's state.

Replacing the symbol X with the symbol E transforms equation (6.2) into the more familiar

$$\Delta E = Q - W, \tag{6.3}$$

7. In this section, the symbols Q and W denote signed quantities. Thus, if $Q > 0$, the system absorbs heat; if $Q < 0$, the system rejects heat; if $W > 0$, the system does work; and if $W < 0$, the system is worked on.

that is, into a standard form of the *first law of thermodynamics,* a statement of principle to which no exceptions have been found. The word *energy,* for which E stands, formed out of Greek stems: *en* meaning "in" or "within" and *ergon* meaning "work," replaces the phrase *work-heat equivalence.*

Note that the first law alone tells us neither what energy is nor how much energy a system contains. Rather, the first law determines only in what way and by how much the energy of a thermodynamic system changes when the system is heated or cooled or does work or is worked on. Occasionally one finds expressions that relate the total energy E of a thermodynamic system to other state variables. However, such expressions are the result of a convention not itself required by the first law of thermodynamics. The first law determines only the ΔE of a process, not the E of a system.

According to the first law, neither heat nor work is contained within a thermodynamic system. Neither does the first law give us reason to affirm the speculations of early natural philosophers (Francis Bacon, Count Rumford, Humphry Davy, Rudolf Clausius, and others—but not Robert von Mayer) that "heat is a form of motion." Rather, *heat* and *work,* Q and W, are names and symbols that describe the kind and quantity of interactions that change a system's state by changing its energy. Given this understanding, many prefer to use the words *heating* and *working* in place of *heat* and *work* whenever possible.[8]

8. See, for instance, Robert Romer, "Heat Is Not a Noun," *American Journal of Physics* 69, no. 2 (2001): 197–109.

6.4 Energy in Thermodynamic and Newtonian Systems

Before the creation of thermodynamics in the mid-nineteenth century, the word *dynamics* invariably referred to *Newtonian dynamics*, which was concerned with deterministic systems composed of a countable number of distinct particles. Such *Newtonian systems* have a total mass, a total momentum, and a total energy, and they can perform or undergo work.

In contrast, a *thermodynamic system* is any part of the universe that, among other state variables, has a temperature that can be changed through heating and cooling. The founders of thermodynamics implicitly, if not explicitly, recognized the existence and importance of thermodynamic systems. While Newtonian systems are useful abstractions, we are surrounded by palpable thermodynamic systems.

Newtonian and thermodynamic systems may perform work on each other and in this way deplete the energy of one and increase the energy of the other. Both possess energy. For this reason, William Thomson and his protégé Peter Guthrie Tait (1831–1901) plausibly characterized physics in the late nineteenth century as "the science of energy."[9]

6.5 The Conservation of Energy

The first law of thermodynamics implies the universal conservation of energy. For if the whole universe is composed of separate, interacting Newtonian and thermodynamic systems, each containing energy, one can in principle follow a bit of energy as it propagates from one system to another, say, from the sun to a

9. Crosbie Smith, *The Science of Energy* (Oxford: Oxford University Press, 1998), 1–2.

plant on the surface of the Earth and then to a human being who consumes the plant and uses it to maintain his or her body temperature. In this way, energy is transported from one place to another. But in no corner of the universe does energy appear out of nowhere or disappear into nothingness.

The suggestion that energy may not always be conserved is usually taken as a sign that something has been overlooked. Consider, for instance, Niels Bohr, who in 1930, briefly and famously, proposed that energy was conserved only statistically, that is, on average, in nuclear beta decays. But then Wolfgang Pauli proposed that a particle, not yet detected and later called the *neutrino*, carried away the apparently missing energy in these decays—a proposal that was eventually verified. Today one can reasonably say, as did Clausius in 1865, that "the energy of the universe is constant."[10]

Each of Clausius's nine memoirs gathered in The Mechanical Theory of Heat *is introduced with a summary of what was then known and what is now to be demonstrated. This excerpt and those that follow in chapters 7, 9, and 11 draw heavily on these summaries.*

In his first memoir, Clausius critiques the material theory of heat that Carnot had adopted. Here Clausius is guided by the notion that heat is a form of motion. This concept, even if not fully justified by what was then known, gave Clausius the freedom to propose that heat was consumed when producing its work equivalent.—DSL

10. W. F. Magie, *A Source Book in Physics* (Cambridge, MA: Harvard University Press, 1963, 1935), 236.

Rudolf Clausius, 1850

"First Memoir: On the Moving Force of Heat and the Laws of Heat Which May Be Deduced Therefrom,"* in his *The Mechanical Theory of Heat* (London: John Van Voorst, 1867), 14–17. Translated from the German by Professor Tyndall in *Philosophical Magazine* 2 (July 1851): 1–21, 102–119.

The steam engine having furnished us with a means of converting heat into a motive power, and our thoughts being thereby led to regard a certain quantity of work as an equivalent for the amount of heat expended in its production, the idea of establishing theoretically some fixed relation between a quantity of heat and the quantity of work which it can possibly produce, from which relation conclusions regarding the nature of heat itself might be deduced, naturally presents itself. Already, indeed, have many successful efforts been made with this view; I believe, however, that they have not exhausted the subject, but that, on the contrary, it merits the continued attention of physicists; partly because weighty objections lie in the way of the conclusions already drawn, and partly because other conclusions, which might render efficient aid towards establishing and completing the theory of heat, remain either entirely unnoticed or have not as yet found sufficiently distinct expression.

The most important investigation in connection with this subject is that of S. Carnot.†

*Communicated in the Academy of Berlin, February 1850, and published in *Poggendorf's Annalen*, March–April 1850, vol. lxxix. pp. 368–397, 500–524.

†*Reflexions stir la puissance motrice du feu, et sur les machines propres a developer cette puissance*, par S. Carnot. Paris, 1824. I have not been able to procure a copy of this work; I know it solely through the writings of Clapeyron and Thomson, from which latter are taken the passages hereafter cited.

Later still the ideas of this author have been represented analytically in a very able manner by Clapeyron.‡

Carnot proves that whenever work is produced by heat and a permanent alteration of the body in action does not at the same time take place, a certain quantity of heat passes from a warm body to a cold one; for example, the vapour which is generated in the boiler of a steam-engine, and passes thence to the condenser where it is precipitated, carries heat from the fireplace to the condenser. This *transmission* Carnot regards as the change of heat corresponding to the work produced. He says expressly, that *no heat is lost* in the process, that the quantity remains unchanged; and he adds, "This is a fact which has never been disputed; it is first assumed without investigation, and then confirmed by various calorimetric experiments. To deny it, would be to reject the entire theory of heat, of which it forms the principal foundation."

I am not, however, sure that the assertion, that in the production of work a loss of heat never occurs, is sufficiently established by experiment. Perhaps the contrary might be asserted with greater justice; that although no such loss may have been directly proved, still other facts render it exceedingly probable that a loss occurs. If we assume that heat, like matter, cannot be lessened in quantity, we must also assume that it cannot be increased; but it is almost impossible to explain the ascension of temperature brought about by friction otherwise than by assuming an actual increase of heat. The careful experiments of Joule, who developed heat in various ways by the application of mechanical force, establish almost to a certainty, not only the possibility of increasing the quantity of heat, but also the fact that the newly-produced heat is proportional to the work expended in its production. It may be remarked further, that many facts have lately transpired which tend to overthrow the hypothesis that heat is itself a body, and to prove that it consists in a motion of the ultimate particles of bodies. If this be so, the general principles of mechanics may be applied to heat; this motion may be converted into

‡*Journal de l'Ecole Polytechnique*, vol. xiv. 1834; *Pogg. Ann.* vol. lix.; and *Taylor's Scientific Memoirs*, Part III. p. 347.

work, the loss of *vis viva* in each particular case being proportional to the quantity of work produced.

These circumstances, of which Carnot was also well aware, and the importance of which he expressly admitted, pressingly demand a comparison between heat and work, to be undertaken with reference to the divergent assumption that the production of work is not only due to an alteration in the *distribution* of heat, but to an actual *consumption* thereof; and inversely, that by the expenditure of work heat may be *produced*. ...

The difference between the two ways of regarding the subject has been seized with much greater clearness by W. Thomson, who has applied the recent investigations of Regnault, on the tension and latent heat of steam, to the completing of the memoir of Carnot.* Thomson mentions distinctly the obstacles which lie in the way of an unconditional acceptance of Carnot's theory, referring particularly to the investigations of Joule, and dwelling on one principal objection to which the theory is liable. If it be even granted that the production of work, where the body in action remains in the same state after the production as before, is in all cases accompanied by a transmission of heat from a warm body to a cold one, it does not follow that by every such transmission work is produced, for the heat may be carried over by simple conduction; and in all such cases, if the transmission alone were the true equivalent of the work performed, an absolute loss of mechanical force must take place in nature, which is hardly conceivable. Notwithstanding this, however, he arrives at the conclusion, that in the present state of science the principle assumed by Carnot is the most probable foundation for an investigation on the moving force of heat. He says, "If we forsake this principle, we stumble immediately on innumerable other difficulties, which, without further experimental investigations, and an entirely new erection of the theory of heat, are altogether insurmountable."

I believe, nevertheless, that we ought not to suffer ourselves to be daunted by these difficulties; but that, on the contrary, we must look

**Transactions of the Royal Society of Edinburgh*, vol. xvi.

steadfastly into this theory which calls heat a motion, as in this way alone can we arrive at the means of establishing it or refuting it. Besides this, I do not imagine that the difficulties are so great as Thomson considers them to be; for although a certain alteration in our way of regarding the subject is necessary, still I find that this is in no case contradicted by *proved facts*. It is not even requisite to cast the theory of Carnot overboard; a thing difficult to be resolved upon, in as much as experience to a certain extent has shown a surprising coincidence therewith. On a nearer view of the case, we find that the new theory is opposed, not to the real fundamental principle of Carnot, but to the addition "no heat is lost"; for it is quite possible that in the production of work both may take place at the same time; a certain portion of heat may be consumed, and a further portion transmitted from a warm body to a cold one; and both portions may stand in a certain definite relation to the quantity of work produced. This will be made plainer as we proceed; and it will be moreover shown, that the inferences to be drawn from both assumptions may not only exist together, but that they mutually support each other.

In the following selection from his fourth memoir, Clausius argues that a state variable exists, here symbolized by U (and in the main text first by X and then by E), that measures the "work-heat equivalence," that is, the energy of a system. Later, in 1864, Clausius explicitly adopted the word energy.*—DSL*

Rudolf Clausius, 1854

"Fourth Memoir: On a Modified Form of the Second Fundamental Theorem in the Mechanical Theory of Heat,"[†]

[†]*Poggendoff's Annalen*, December 1854, Vol. xciii. p. 481; translated in the *Journal de Mathématiques*, Vol. xx. Paris, 1855, and in the *Philosophical Magazine*, August 1856, S. 4. Vol. xii. p. 81.

in his *The Mathematical Theory of Heat* (London: John Van Voorst, 1867), 112–113.

Theorem of the Equivalence of Heat and Work

Whenever a moving force generated by heat acts against another force, and motion in the one direction or the other ensues, positive work is performed by the one force at the same time that negative work is done by the other. As this work has only to be considered as a simple quantity in calculation, it is perfectly arbitrary, in determining its sign, which of the two forces is chosen as the indicator. Accordingly in researches which have a special reference to the moving force of heat, it is customary to determine the sign by counting as positive the work done by heat in overcoming any other force, and as negative the work done by such other force. In this manner the theorem of the equivalence of heat and work, which forms only a particular case of the general relation between *vis viva* and mechanical work, can be briefly enunciated thus:

Mechanical work may be transformed into heat, and conversely heat into work, the magnitude of the one being always proportional to that of the other.

The forces which here enter into consideration may be divided into two classes: those which the atoms of a body exert upon each other, and which depend, of course, upon the nature of the body, and those which arise from the foreign influences to which the body may be exposed. According to these two classes of forces which have to be overcome (of which the latter are subjected to essentially different laws), I have divided the work done by heat into interior and exterior work.

With respect to the interior work, it is easy to see that when a body, departing from its initial condition, suffers a series of modifications and ultimately returns to its original state, the quantities of interior work thereby produced must exactly cancel one another. For if any positive or negative quantity of interior work had remained, it must have produced an opposite exterior quantity of work or a change in the existing quantity of heat; and as the same process could be repeated any number of times, it would be possible, according to the sign, either to produce work or heat continually from nothing, or else to lose work or heat continually, without obtaining any equivalent; both of which cases are universally allowed to be impossible. But if at every return of the body to its initial condition the quantity of interior work is zero, it follows, further, that the interior work corresponding to any given change in the condition of the body is completely determined by the initial and final conditions of the latter, and is independent of the path pursued in passing from one condition to the other. Conceive a body to pass successively in different ways from the first to the second condition, but always to return in the same manner to its initial state. It is evident that the quantities of interior work produced along the different paths must all cancel the common quantity produced during the return, and consequently must be equal to each other.

It is otherwise with the exterior work. With the same initial and final conditions, this can vary just as much as the exterior influences to which the body may be exposed can differ.

Let us now consider at once the interior and exterior work produced during any given change of condition. If opposite in sign they may partially cancel each other, and what remains must then be proportional to the simultaneous change which has occurred in the quantity of existing heat. In calculation, however, it amounts to the same thing if we assume an alteration in the quantity of heat equivalent to each of the two kinds of work. Let Q, therefore, be the quantity of heat which must be imparted to a body during its passage, in a given manner, from one condition to another, any heat withdrawn from the body being counted as an imparted negative quantity of heat. Then Q may be divided into three parts, of which the first is employed in increasing the heat actually existing in the body, the second in producing the interior, and the third in producing the exterior work. What was before stated of the second part also applies to the first—it is independent of the path pursued in the passage of the body from one state to another: hence both parts together may be represented by one function U, which we know to be completely determined by the initial and final states of the body. The third part, however, the equivalent of exterior work, can, like this work itself, only be determined when the precise manner in which the changes of condition took place is known. If W be the quantity of exterior work, and A the equivalent of heat for the unit of work, the value of the third part will be $A \cdot W$, and the first fundamental theorem will be expressed by the equation $Q = U + A \cdot W$.

7
Second Law of Thermodynamics

7.1 The Independence of the First and Second Laws

Rudolf Clausius identified two foundational principles in "On the Motive Power of Heat and on the Laws which can be deduced from it for the Theory of Heat" (1850).[1] His "first principle," which developed into the first law of thermodynamics, was discovered, verified, and articulated in different ways by Robert von Mayer, James Joule, L. A. Colding, and Hermann von Helmholtz in the period 1842 to 1847.[2] The second principle, "the essential principle of Carnot," is that,

1. Rudolf Clausius, "On the Motive Power of Heat and the Laws Which Can Be Deduced from It for the Theory of Heat," in *Reflections on the Motive Power of Fire and Other Papers on the Second Law of Thermodynamics* (Gloucester, MA: Peter Smith, 1977, 1960), 109–152.
2. See Hermann von Helmholtz, "The Conservation of Energy," in W. F. Magie, *A Source Book in Physics* (Cambridge, MA: Harvard University Press, 1963, 1935), 212–220. See also Thomas Kuhn, "Energy Conservation as an Example of Simultaneous Discovery," in *Critical Problems in the History of Science*, edited by Marshal Clagett (Madison: University of Wisconsin Press, 1959), 321–356. Of the twelve people Thomas Kuhn claimed could be credited with originating the first law of thermodynamics, Colding, Mayer, Helmholtz, and Joule stand out.

> Whenever work is done by heat and no permanent change occurs in the condition of the working body, a certain quantity of heat passes from a hotter to a colder body. (1850)

This statement echoes similar ones Carnot made in the *Reflections* (1824) that we take as the earliest statements of the second law of thermodynamics. Interestingly, Clausius's ordering of these two principles or laws, for which he gives no justification, reverses the order of their discovery.

That the content of Clausius's first principle is independent of the content of his second principle is never doubted. By *independent,* I mean that one can discuss the mechanical equivalent of heat, the first law of thermodynamics, and the conservation of energy without mentioning or assuming the truth of the second principle. But that the second principle, that is, the second law of thermodynamics, is similarly independent of the first principle or law is almost never acknowledged.[3]

That the original statements of the first and second laws of thermodynamics are indeed independent of one another follows from their history. Clausius, after all, identifies his "second principle" with "the essential principle of Carnot" and Carnot worked without benefit of the first law. Therefore, the truth of Carnot's "essential principle" must be independent of the truth of the first law.

The independence of the first and second laws is also logically appealing. Many statements advertised as versions of the second law of thermodynamics, such as the law of increase of entropy, are actually deductions from the first law and an independent

3. For an exception, see H. W. Schamp Jr., "Independence of the First and Second Laws of Thermodynamics," *American Journal of Physics* 30 (1962): 825–829.

version of the second law. Therefore, the deduction produced is logically as much a product of the first law as of the second.

Versions of the second law that are independent of the truth of the first law of thermodynamics can be framed in various ways. Each of the ways we examine here is sufficient as a basis on which to found classical thermodynamics. Yet not all are logically equivalent. None of these early, independent versions of the second law depend on the concept of reversibility, Thomson's definition of absolute temperature, or the concept of entropy.

7.2 Carnot's Second Law

That *no heat engine can produce work without transporting heat from a reservoir at a higher temperature to one at a lower temperature* is, in section 3.2, referred to as *Carnot's law*. Here we call this statement *Carnot's version of the second law.*[4]

Of course, Carnot also believed that a heat engine loses no heat in its transport from a hotter to a colder body. Stripped of this inessential restriction, identified by Clausius as a "subsidiary statement," Carnot's second law can be expressed in this way: *In order to produce work, a heat engine must transport heat between at least two heat reservoirs.*

One can also express the second law in terms of *Carnot's simplest heat engine*: a heat engine, illustrated in figure 7.1, that in one cycle receives heat Q_H from a hot reservoir at temperature t_H, produces work W, and gives up heat Q_C (not necessarily equal to either Q_H or to $Q_H - W$) to a colder reservoir at temperature

4. D. S. Lemons and M. Penner, "Sadi Carnot's Contribution to the Second Law of Thermodynamics," *American Journal of Physics* 76 (2008): 21–25.

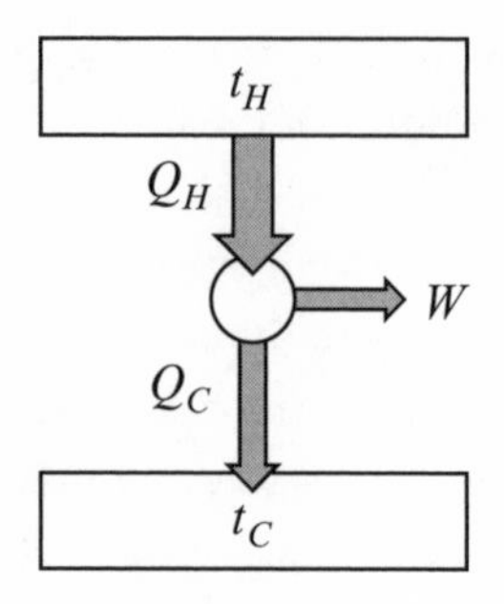

Figure 7.1
Carnot's simplest heat engine.

t_C. In this way, Carnot's second law becomes the statement that *no heat engine operating in a cycle can be simpler than Carnot's simplest heat engine.*

Since eliminating heat reservoirs simplifies a heat engine, the "heat engines," or cyclic processes, thus prohibited, are each marked in figure 7.2 with the symbol ⦸, meaning "prohibited." These processes are too simple.

Today we find the prohibitions illustrated in figures 7.2a and 7.2c unnecessary as each violates the first law of thermodynamics. In 1824, Carnot would have found the prohibitions illustrated in figures 7.2b and 7.2c unnecessary since neither conserves caloric. For this reason, Carnot focused on the absurdity of figure 7.2a. All three prohibitions illustrated in figure 7.2 express Carnot's second law independent of the conservation of any quantity.

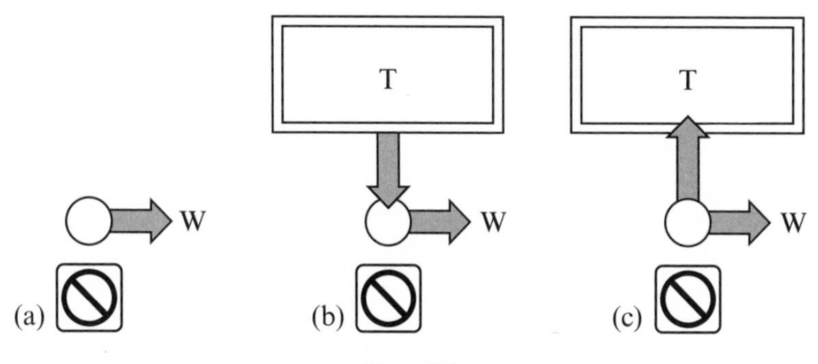

Figure 7.2
"Heat engines" or cyclic processes simpler than Carnot's simplest heat engine that are prohibited by Carnot's version of the second law of thermodynamics.

7.3 Clausius's Second Law

In the process of disentangling Carnot's essential principle from its inessential restriction, Clausius fashioned his own version of the second law in 1850. A few months later, in 1851, William Thomson proposed a different version of it. Clausius and Thomson each conceived of their versions of the second law as axioms that although motivated by empirical evidence need no proof.[5] Their statements of the second law, while verbally and logically distinct, are each sufficient, along with the first law, for the purpose of founding the science of thermodynamics. And each can

5. Clausius, "On the Motive Power of Heat," 134; William Thomson, "Dynamical Theory of Heat," in *Transactions of the Royal Society of Edinburgh* (March 1851), in *Mathematical and Physical Papers* (Cambridge: Cambridge University Press, 1882), 174–323, and in an excerpt in Magie, *Source Book in Physics*, 244–246.

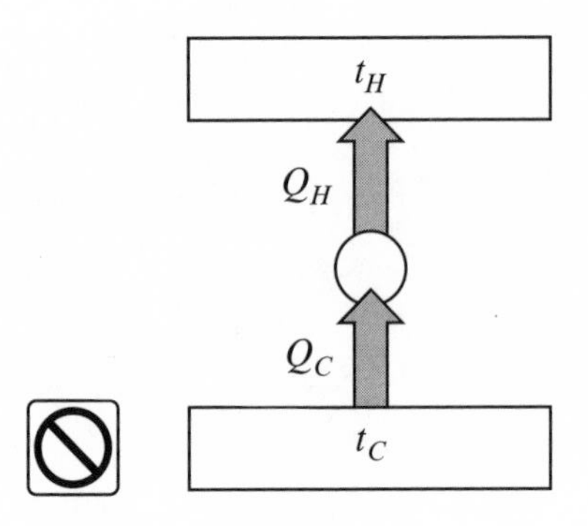

Figure 7.3
Heat process prohibited by Clausius's version of the second law of thermodynamics.

be framed in a way that does not depend on the truth of the first law.

According to Clausius's second law, *no process is possible whose only result is to transfer heat from a colder reservoir to a hotter one.* Thus, the terms of Clausius's second law are those of everyday phenomena. The hot coffee on my desk always cools down and never, by extracting energy from its colder environment, becomes hotter. Figure 7.3, in which heat Q_C is extracted from a colder reservoir at temperature t_C and heat Q_H, not necessarily equal to Q_C, is received by a hotter reservoir at temperature t_H, illustrates the heat process prohibited by Clausius's second law.

7.4 Thomson's Second Law

Thomson recognized that "the whole theory of the motive power of heat is founded on ... two propositions, due respec-

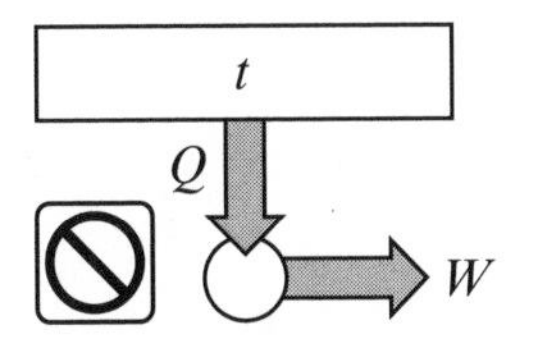

Figure 7.4
Heat engine prohibited by Thomson's second law.

tively to Joule [the first law], and to Carnot and Clausius [the second law]," and had "no wish to claim [a] priority ... that is entirely due to Clausius."[6] Even so, Thomson explored the subject independently.

According to Thomson's version of the second law, *no heat engine, operating in a cycle, can produce work by extracting heat from a single heat reservoir.* Figure 7.4 illustrates the process Thomson's second law prohibits. Here a heat engine, during one cycle of its operation, receives heat Q from a reservoir at temperature t and produces work W, not necessarily equal to Q.

Thomson's second law restricts the world less severely than does Carnot's, as a comparison of figures 7.2 and 7.4 reveals. Yet both are sufficient as axioms on which to found thermodynamics. After all, the first law makes the difference between the two irrelevant. However, Clausius's second law, as can be shown, is logically equivalent to Carnot's second law.[7] All three of these versions of the second law are logically equivalent once the truth of the first law is assumed.

6. Thomson, "Dynamical Theory of Heat," in Magie, *A Source Book in Physics*, 246.
7. Don S. Lemons, *Mere Thermodynamics* (Baltimore: Johns Hopkins University Press, 2009), chap. 5.

7.5 "As Many Formulations"

Which version of the second law of thermodynamics one prefers is a matter of taste and the use to which it is put. My own sense is that Clausius's second law most strongly captures everyday experience, Thomson's second law, when coupled with the first law, reflects the impossibility of heat engines that are 100 percent efficient, and Carnot's, being the first expression of the second law, is historically important.

According to Nobel laureate Percy Bridgman, "There have been nearly as many formulations of the second law as there are discussions of it."[8] One of the most popular and succinct expressions of the second law is that *the entropy of an isolated system never decreases*. But the concept of entropy is as much a consequence of the first law as it is of an independent version of the second, and for this reason it is not properly associated with only one of the two laws of classical thermodynamics.

I have, in contrast, taken pains to formulate Carnot's, Clausius's, and Thomson's versions of the second law in such a way that their truth does not depend on the truth of the first law. Just as one can draw useful conclusions from *only* the first law, the second law *alone* also has certain interesting consequences.[9] Indeed, Carnot's discovery of some of these consequences, as outlined in chapter 3, has made him justly famous. Of course, the two laws together compose the foundation of classical thermodynamics.

8. P. W. Bridgman, *The Nature of Thermodynamics* (Cambridge, MA: Harvard University Press, 1941), 116.

9. Two textbooks, J. T. Vanderslice and H. W. Schamp Jr., *Thermodynamics* (Englewood Cliffs, NJ: Prentice Hall, 1966), and Lemons, *Mere Thermodynamics*, explore the consequences of an independent version of the second law apart from the truth of the first law.

Here Clausius searches for an axiom sufficient to derive the theorems of classical thermodynamics including, most importantly, Carnot's theorem. The selection concludes with Clausius adopting a version of the second law according to which "heat cannot of itself pass from a colder to a warmer body."—DSL

Rudolf Clausius, 1863

"Seventh Memoir: On an Axiom in the Mechanical Theory of Heat,"* in his *The Mechanical Theory of Heat* (London: John Van Voorst, 1867), 267–270.

1. When I wrote my First Memoir on the *Mechanical Theory of Heat*, two different views were entertained relative to the deportment of heat in the production of mechanical work. One was based on the old and widely spread notion, that heat is a peculiar substance, which may be present in greater or less quantity in a body, and thus determine the variations of temperature. Conformably with this notion was the opinion that, although heat could change its mode of distribution by passing from one body into another, and could further exist in different conditions, to which the terms *latent* and *free* were applied, yet the quantity of heat in the whole mass could neither increase nor diminish, since matter can neither be created nor destroyed.

Upon this view is based the paper published by S. Carnot, in the year 1824, wherein machines driven by heat are subjected to a general theoretical treatment. Carnot, in investigating more closely the circumstances under which moving force can be produced by heat, found that in all cases there is a passage of heat from a body of higher into one of a lower temperature; as in the

*Read at a Meeting of the Swiss Association, held at Samaden, August 25th, 1863, and published in *Poggendorft's Annalen*, November 1863, vol. CXX. p. 426.

case of a steam-engine where, by means of steam, heat passes from the fire or from a body of very high temperature, to the condenser, a space containing bodies of lower temperature. He compared this manner of producing work with that which occurs when a mass of water falls from a higher to a lower level, and consequently, in correspondence with the expression *une chute d'eau* [waterfall], he described the fall of heat from a higher to a lower temperature as *une chute du calorique.*

Regarding the subject from this point of view, he lays down the theorem that the magnitude of the work produced always bears a certain general relation to the simultaneous transfer of heat, i. e. to the quantity of heat which passes over, and to the temperatures of the bodies between which the transfer takes place, and that this relation is independent of the nature of the substances through which the production of work and the transfer of heat are effected. His proof of the necessity of such a relation is based on the axiom that *it is impossible to create a moving force out of nothing,* or in other words, *that perpetual motion is impossible.*

The other view above referred to is that heat is not invariable in quantity; but that when mechanical work is produced by heat, heat must be consumed, and that, on the contrary, by the expenditure of work a corresponding quantity of heat can be produced. This view stands in immediate connection with the new theory respecting the nature of heat, according to which heat is not a substance but a motion. Since the end of the last century various writers, amongst whom Rumford, Davy, and Seguin may be mentioned, have accepted this theory; but it is only since 1842 that Mayer of Heilbronn, Colding of Copenhagen, and Joule of Manchester examined the theory more closely, founded it, and established with certainty the law of the equivalence of heat and work.

According to this theory, the causal relation involved in the process of the production of work by heat is quite different from that which Carnot assumed. Mechanical work ensues from the conversion of existing heat into work, just in the same manner as, by the ordinary laws of mechanics, force is overcome, and work thereby produced, by motion which already exists; in the latter case the motion suffers a loss, in *vis viva*, equivalent to the work done, so that we may say that the *vis viva* of motion has been converted into work. Carnot's comparison, therefore, in accordance with which the production of work by heat corresponds to the production of work by the falling of a mass of water,—and, in fact, the fall of a certain quantity of heat from a higher to a lower temperature may be regarded as a cause of the work produced, was no longer admissible according to modern views. On this account it was thought that one of two alternatives must necessarily be accepted; either Carnot's theory must be retained and the modern view rejected, according to which heat is consumed in the production of work, or, on the contrary, Carnot's theory must be rejected and the modern view adopted.

2. When at the same period I entered on the investigation of this subject, I did not hesitate to accept the view that heat must be consumed in order to produce work. Nevertheless I did not think that Carnot's theory, which had found in Clapeyron a very expert analytical expositor, required total rejection; on the contrary, it appeared to me that the theorem established by Carnot, after separating one part and properly formulizing the rest, might be brought into accordance with the more modern law of the equivalence of heat and work, and thus be employed together with it for the deduction of important conclusions. The theorem of Carnot thus modified was treated by me in the second part of the above-cited memoir, in the first part of which I had considered the law of the equivalence of heat and work.

In my later memoirs I succeeded in establishing simpler and at the same time more comprehensive theorems by pursuing further the same considerations which had led me to the first modification of Carnot's theorem. I will not now enter, however, upon these extensions of the theory, but will limit myself for the present to the question how, in accordance with the law of the equivalence of heat and work, the necessity can be demonstrated of the other theorem in its modified form.

The axiom employed by Carnot in the proof of his theorem, and which consists in the impossibility of creating moving force, or, more properly expressed, mechanical work out of nothing, could no longer be employed in establishing the modified theorem. In fact, since in the latter it is already assumed that to produce mechanical work an equivalent amount of heat must be consumed, it follows that the supposition of the creation of work is altogether out of the question, no matter whether a transfer of heat from a warm to a colder body does or does not accompany the consumption of heat.

On the other hand, I found that another and, in my opinion, a more certain basis can be secured for the proof by reversing the sequence of reasoning pursued by Carnot, and by accepting as an axiom a theorem, in a somewhat modified form, which may be regarded as a consequence of his assumptions.

In fact, after establishing from the axiom that work cannot be produced from nothing, the theorem that in order to produce work a corresponding quantity of heat must be transferred from a warmer to a colder body, Carnot to be consistent could not but conclude that, in order to transfer heat from a colder to a warmer body, work must be expended. Although we must now abandon the argument which led to this result, and not withstanding the fact that the result itself in its original form is not quite admissible, it is nevertheless manifest that an essential difference exists between the transfer of heat from a warmer to a colder body and the transfer from a colder to a warmer, since the first may take place spontaneously under circumstances which render the latter impossible.

On investigating the subject more closely, and taking into consideration the known properties and actions of heat, I came to the conviction that the difference in question had its origin in the nature of heat itself, inasmuch as by its very nature it must tend to equalize existing differences of temperature. Heat accordingly incessantly strives to pass from warmer to colder bodies, and a passage in a contrary direction can only take place under circumstances where simultaneously another quantity of heat passes from a warmer to a colder body, or when some change occurs which has the peculiarity of not being reversible without causing on its part such a transfer from a warmer to a colder body. This change which simultaneously takes place is consequently to be regarded as the equivalent of that transfer of heat from a colder to a warmer body, so that it cannot be said that the transfer has taken place of itself (*von selbst*).

I thought it permissible, therefore, to lay down the axiom, that *Heat cannot of itself pass from a colder to a warmer body*, and to employ it in demonstrating the second fundamental theorem of the mechanical theory of heat.

8
Absolute Temperature—Again

8.1 Another Try

Rudolf Clausius and William Thomson's new understanding of thermodynamics undermined Thomson's first definition of absolute temperature. Recall that in 1848, Thomson proposed a definition of absolute temperature independent of any one substance by adopting the working fluid of a reversible Carnot engine as its thermometric fluid and the work that fluid produced, during one cycle, as its thermometric variable. Thomson's 1848 definition depended on connecting reversible heat engines in series with each engine receiving the heat rejected by the engine upstream and passing on the same quantity of heat to the engine downstream, as illustrated in figure 4.1. Therefore, this scheme would appear to fall apart if a heat engine does not pass on the same heat it receives.

Thomson's response was to retain the concept of an absolute temperature based on the working fluid of a reversible Carnot engine, but to change the way in which that temperature is related to its transformations. He reasoned that

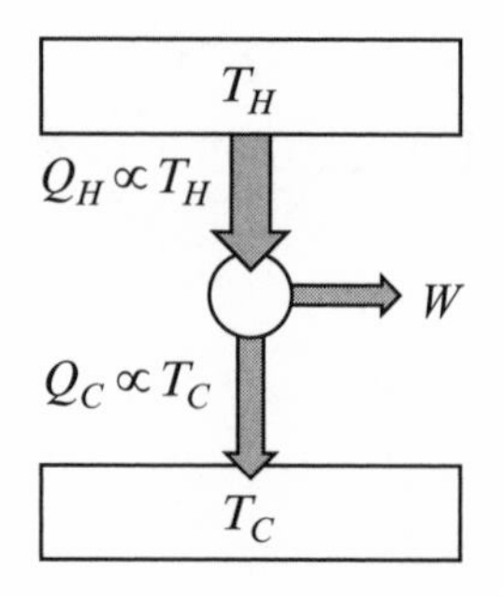

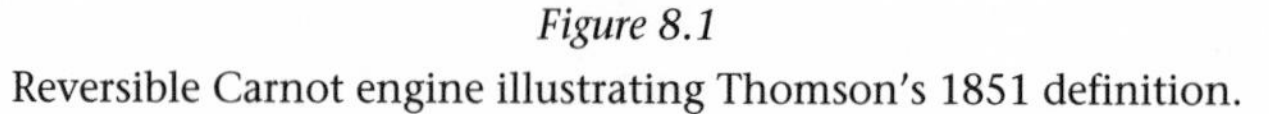
Figure 8.1
Reversible Carnot engine illustrating Thomson's 1851 definition.

> If two bodies are put in contact, and neither gives heat to the other, their temperatures are said to be the same; but if one gives heat to the other, its temperature is said to be higher. (1851)[1]

Therefore (in Thomson's typically wordy style),

> The temperatures of two bodies are proportional to the quantities of heat taken in and given out in localities at one temperature and at the other, respectively by a material system subjected to a complete cycle of perfectly reversible thermodynamic operations, and not allowed to part with or take in heat at any other temperature: or the absolute values of two temperatures are to one another in the proportion of the heat taken in to the heat rejected in a perfect thermo-dynamic engine working with a source and refrigerator at the higher and lower of the temperatures respectively. (1851)

Figure 8.1 of a reversible Carnot engine that in one cycle receives heat Q_H from its source, produces work W, and rejects

1. W. Thomson, "Dynamical Theory of Heat," in *Transactions of the Royal Society of Edinburgh* (1851) and article 48 in *Physical and Mathematical Papers* (Cambridge: Cambridge University Press, 1882), 235.

heat Q_C to its sink illustrates Thomson's new definition of absolute temperature. Accordingly, either $T_H \propto Q_H$ and $T_C \propto Q_C$ or, equivalently,

$$\frac{T_C}{T_H} = \left(\frac{Q_C}{Q_H}\right). \tag{8.1}$$

Note that equation (8.1) defines only the ratio T_C/T_H and that a complete determination of the two temperatures, T_C and T_H, requires a second condition. For this second condition, Thomson chose to specify the temperature difference $T_H - T_C$ in a way that conformed to current use:

> Two fixed points of temperature being chosen according to Sir Isaac Newton's suggestions, by particular effects on a particular substance or substances, the difference of these temperatures is to be called unity, or any number of units or degrees as may be found convenient. The particular convention is, that the difference of temperatures between the freezing- and boiling-points of water under standard atmospheric pressure shall be called 100 degrees. (1851)

Therefore, if T_B and T_F are, respectively, the absolute temperatures assigned to the standard boiling and freezing points of water, then

$$T_B - T_F = 100°. \tag{8.2}$$

Degrees of temperature defined according to equations (8.1) and (8.2) are now known as *degrees Kelvin*. (Thomson was made Baron Kelvin of Largs in 1892.)

Interestingly, equation (8.2), based as it is on the properties of water, privileges this one material over all others. Another "water-based, absolute scale" is William Rankine's: he adopted equation (8.1) but replaced equation (8.2) with

$$T_B - T_F = 180° \tag{8.3}$$

so that intervals on this scale coincide with intervals of Fahrenheit degrees. Degrees of temperature defined according to equations (8.1) and (8.3) are known as *degrees Rankine*. Any temperature scale based on equation (8.1) is an *absolute scale*.

8.2 Thomson's Simple Choice

Stipulative definitions are inherently arbitrary. The arbitrariness encapsulated in equations (8.2) and (8.3) is obvious, but equation (8.1), that is, $T_C/T_H = Q_C/Q_H$, also represents only one of a number of choices Thomson could have made.

Recall that the efficiency of a reversible Carnot engine in terms of the heat received Q_H and the work produced W $[= Q_H - Q_C]$ in one cycle is given by

$$\begin{aligned} \varepsilon(t_H, t_C) &= \frac{W}{Q_H} \\ &= \frac{Q_H - Q_C}{Q_H} \\ &= 1 - \frac{Q_C}{Q_H}, \end{aligned} \tag{8.4}$$

where the indicated dependence of $\varepsilon(t_H, t_C)$ is on the empirical temperatures t_H and t_C, respectively, of the hot and cold reservoirs between which the engine operates. This dependence expresses Carnot's deduction that *the efficiency of a reversible engine depends only on the empirical temperatures, t_H and t_C, of the heat reservoirs between which the engine operates*. Given equation (8.4), the heat ratio Q_C/Q_H also depends only on t_H and t_C. These relationships gave Thomson permission to define an absolute temperature scale by stipulating a relation between the heat ratio Q_C/Q_H of a Carnot engine and the ratio T_C/T_H of absolute temperatures.

Carnot's analysis of a reversible heat engine, as recapitulated in section 3.6, obliged Thomson to build into his definition of absolute temperature only two conditions: (1) when $t_H = t_C$, and thus when $T_H = T_C$, the efficiency $\varepsilon = 0$ and so $Q_H/Q_C = 1$, and (2) when t_H is fixed (and so also is T_H), the efficiency ε decreases as t_C (and so also as T_C) increases. Thomson could, for instance, have met these requirements by replacing equation (8.1) with the relation

$$\frac{T_C}{T_H} = \left(\frac{Q_C}{Q_H}\right)^n, \tag{8.5}$$

where n is any positive number.

Thomson gave no justification for the choice he did make, $n = 1$, although his unremarked justification might have been its simplicity. The simple, if arbitrary, choice $n = 1$ leading to equation (8.1), that is, to $Q_C/Q_H = T_C/T_H$, also leads to simple thermodynamic relationships and simple equations of state. For example, Thomson's choice leads to a simple expression for the efficiency of a reversible Carnot engine,

$$\frac{W}{Q_H} = 1 - \frac{T_C}{T_H}, \tag{8.6}$$

and also to the simple equation of state $PV = nRT$ for the pressure P of an ideal gas in terms of its volume V, mole number n, and absolute temperature T.

8.3 Absolute Zero

Does the adjective *absolute* in "absolute zero" refer to the zero of an absolute temperature scale, or does *absolute* mean "the lowest possible temperature"? Today these two meanings converge, but they did not always. One sense of absolute zero emerged in the eighteenth century as a corollary to the doctrine of caloric. Accordingly, the temperature of a body containing no caloric was thought to be absolute zero. Eighteenth-century estimates of absolute zero varied wildly.[2]

Relations among the pressure, volume, and temperature of a gas, as they were understood in the early nineteenth century, suggested yet another meaning. Absolute zero was that temperature at which the product of the pressure and volume of a gas vanished. Given the empirical temperature scales in use at the time, this version of absolute zero was a negative number. For example, according to the determination of John Herapath (1790–1868), absolute zero is –448° F (–267° C). However, in 1821, an unknown writer offered the relevant observation that a gas might cease to be gaseous, and thus render the ideal gas law useless, at a temperature much above absolute zero.[3]

Thomson's absolute temperature scale, based as it is on equation (8.1), gives absolute zero its modern meaning. According to equation (8.6), absolute zero is the temperature of the cold reservoir that makes a reversible Carnot engine 100 percent efficient. But no heat engine can be 100 percent efficient, much less more than 100 percent efficient. A 100 percent efficient heat engine

2. See references to "temperature, absolute zero of" in the index of D. S. L. Cardwell, *From Watt to Clausius: The Rise of Thermodynamics in the Early Industrial Age* (Ithaca, NY: Cornell University Press, 1971).
3. Ibid., 148.

would produce no waste heat Q_C, and that would violate Thomson's version of the second law. Therefore, no absolute temperature can be equal to or lower than zero.[4]

8.4 The 1848 and 1851 Definitions Harmonized

Thomson appeared not to have realized that he incorporated into his 1851 definition an important feature of his earlier 1848 definition. A close look at Thomson's 1848 definition of absolute temperature, as recapitulated in sections 4.2 and 4.3, reveals only that whatever passes through a reversible heat engine must be *conserved* and must also be *independent of the temperature interval* of the reservoirs between which the Carnot engine operates. The only such quantity is Q_H/T_H. And given Thomson's 1851 definition $Q_C/T_H = Q_C/T_C$, we are free to imagine that while Q_H/T_H passes from the hotter reservoir into the working fluid, Q_C/T_C passes from the working fluid into the colder reservoir.

Thus, a series of Carnot engines connected in series does indeed pass on, from one heat reservoir to the other, a conserved quantity that is independent of temperature interval. That quantity is the heat transferred reversibly to and from a heat reservoir divided by the absolute temperature of that reservoir. A few years later, in 1854, Rudolf Clausius discovered the necessity and purpose of this quantity and, in 1865, gave that quantity the name *entropy*.

4. This statement is the case when absolute temperatures are based on Thomson's 1851 definition. An alternative definition based on the energy derivative of the entropy function does, in special circumstances, allow for negative absolute temperatures.

These two paragraphs, in which Thomson offers his second definition of absolute temperature, appear almost as an aside in this 176-page paper.—DSL

William Thomson (Kelvin), 1851

"An Account of Carnot's Theory of the Motive Power of Heat; with Numerical Results Deduced from Regnault's Experiments on Steam," in his *Mathematical and Physical Papers* (London: Cambridge University Press, 1882), 1:235–236.

[From *Transactions of the Edinburgh Royal Society*, xvi. 1851; *Annal. de Chimie*, xxxv. 1852.]

99. *Definition of temperature and general thermometric assumption.* If two bodies be put in contact, and neither gives heat to the other, their temperatures are said to be the same; but if one gives heat to the other, its temperature is said to be higher.

The temperatures of two bodies are proportional to the quantities of heat respectively taken in and given out in localities at one temperature and at the other, respectively, by a material system subjected to a complete cycle of perfectly reversible thermodynamic operations, and not allowed to part with or take in heat at any other temperature: or, the absolute values of two temperatures are to one another in the proportion of the heat taken in to the heat rejected in a perfect thermo-dynamic engine working with a source and refrigerator at the higher and lower of the temperatures respectively.

100. *Convention for thermometric unit, and determination of absolute temperatures affixed points in terms of it.* Two fixed points of temperature being chosen according to Sir Isaac Newton's suggestion, by particular effects on a particular substance or substances,

the difference of these temperatures is to be called unity, or any number of units or degrees as may be found convenient. The particular convention is, that the difference of temperatures between the freezing- and boiling-points of water under standard atmospheric pressure shall be called 100 degrees. The determination of the absolute temperatures of the fixed points is then to be effected by means of observations indicating the economy of a perfect thermo-dynamic engine, with the higher and the lower respectively as the temperatures of its source and refrigerator. The kind of observation best adapted for this object was originated by Mr Joule, whose work in 1844* laid the foundation of the theory, and opened the experimental investigation; and it has been carried out by him, in conjunction with myself, within the last two years, in accordance with the plan proposed in Part IV of the present series. The best result, as regards this determination, which we have yet been able to obtain is, that the temperature of freezing water is 273.7 on the absolute scale; that of the boiling point being consequently 373.7.† Further details regarding the new thermometric system will be found in a joint communication to be made by Mr. Joule and myself to the Royal Society of London before the close of the present session.

*"On the Changes of Temperature occasioned by the Rarefaction and Condensation of Air," see Proceedings of the Royal Society, June 1844; or, for the paper in full. *Phil. Mag.* May 1845.

†[Note of Dec. 1881. Later results show that these numbers are more accurately 273.1 and 373.1. Article on Heat by the author, *Encyc. Brit.;* also published separately under the title "Heat," Edinburgh, Black, 1880.]

9
Entropy

9.1 The Word *Entropy*

Mid-nineteenth-century physicists used the new and unfamiliar words *energy* and *entropy* to name concepts that were themselves new and unfamiliar. In time, *energy* has become a familiar, everyday word, while the denotations of *entropy* remain relatively unfamiliar. This is not because the word *energy* is lexically more appropriate than the word *entropy*; rather, identifying energy in simple Newtonian, one- and two-particle systems has become commonplace, while entropy is found only in relatively complex thermodynamic systems.

Work and heat describe what is done to a system or what the system does, while entropy, like energy, is something a system possesses. Clausius at first called this state variable *equivalence-value, transformation-value*, or *transformation content*:

> But as I hold it to be better to borrow terms for important magnitudes from the ancient languages, so that they may be adopted unchanged in all modern languages, I propose to call the magnitude *S* the *entropy* of the body, from the Greek word τροπη, *transformation*. I have intentionally formed the word *entropy* so as to be as similar as possible to the word *energy*; for the two magnitudes to be denoted by these words are so nearly allied in their physical

> meanings, that a certain similarity in designation appears to be desirable. (1865)[1]

Energy and *entropy* are "so nearly allied in their physical meanings" that much of what has been said of energy can also be said, word for word, of entropy. Adopting the language of section 6.3, classical thermodynamics teaches us neither what entropy is nor how much entropy a system possesses. Rather, classical thermodynamics teaches us only in what way and by how much the entropy of a system can be incremented. Occasionally one finds expressions for the total entropy of a system. However, such expressions are the result of a convention not itself required by the first and second laws of thermodynamics—or even by the third law. Classical thermodynamics determines only the increment of a system's entropy ΔS, not its total quantity S.

9.2 Incrementing Entropy

Recall that according to Thomson's 1851 definition, the absolute temperatures T_H and T_C are related to the heat flows Q_H and Q_C in a reversible heat engine operating cyclically between two heat reservoirs by $Q_H/T_H = Q_C/T_C$. This relation suggests a conservation law,

$$0 = \frac{Q_H}{T_H} - \frac{Q_C}{T_C}, \tag{9.1}$$

that seems to describe how, in a reversible cycle, the working fluid receives a quantity Q_H/T_H from the hotter reservoir and

1. Rudolf Clausius, "On Several Convenient Forms of the Fundamental Equations of the Mechanical Equivalent of Heat," in his *The Mechanical Theory of Heat* (London: Van der Voorst, 1867), 357. This article was originally read at the Philosophical Society of Zurich on April 24, 1865.

rejects the same quantity, styled Q_C/T_C, to the colder reservoir. Alternatively and equivalently, the hotter reservoir rejects a quantity Q_H/T_H, while the colder reservoir receives a quantity Q_C/T_C. The reversibility of these heat flows ensures that in a cyclic two-reservoir heat engine a heat reservoir loses or gains what the working fluid gains or loses.

Thus, quantities of the form Q/T denote increments to the entropy of a system with constant temperature T. Note that if we make the heat transferred to or from a finite object small enough, denoted, say, by δQ, the finite object is effectively a heat reservoir with constant temperature T whose entropy in reversible heating or cooling is incremented by $\delta Q/T$.[2]

9.3 Entropy's Deep Foundations

William Thomson's 1851 definition of absolute temperature, however suggestive, does not prove that the state variable entropy is necessary or that the entropy of a system is incremented by quantities of the form $\delta Q/T$. Rather, it was Rudolf Clausius who, in 1854, showed that the first law of thermodynamics and an independent version of the second law (for instance, that of Carnot, Clausius, or Thomson) require that every thermodynamic system is characterized with a state variable, having certain properties—a state variable we now call *entropy*.[3]

2. The notation for an indefinitely small quantity that is not an integrable differential, such as δQ, δW, $đQ$, or $đW$, is not as widely recognized as that for an integrable differential like dE or dS.
3. Clausius's proof is contained in the fourth (1854) memoir of the series of papers on thermodynamics in his *Mechanical Theory of Heat.* Also see proofs of Clausius's theorem and its consequences in Kerson Huang, *Statistical Mechanics* (Hoboken, NJ: Wiley, 1966), 15–17; Mark Zemansky, *Heat and Thermodynamics* (New York: McGraw-Hill, 1957), 168–172; and Don S. Lemons, *Mere Thermodynamics* (Baltimore: Johns Hopkins University Press, 2009), 50–54, 57–61.

Clausius's proof follows the pattern of showing that the net increment $\sum \delta Q/T$ to a system in any reversible process that takes the system from one state to another is the same. Here, δQ denotes the *signed* quantity of heat received by the system with temperature T during one stage of the process. Thus, if the system receives heat during this stage $\delta Q > 0$, and if the system rejects heat during this stage $\delta Q < 0$. The symbol Σ indicates that the signed quantities $\delta Q/T$ are summed over every stage of the reversible process.

Clausius's proof showed that the value of $\sum \delta Q/T$ depends only on the identity of the initial and final states of the process and not at all on the particular reversible path taken between the two states. This is equivalent to saying that the net increment $\sum \delta Q/T$ to a system during any reversible process is an increment of a state variable S (the entropy) and that the net increment to the entropy of a system during a complete, reversible cycle vanishes.

Textbooks usually give Clausius's proof of this theorem neither in detail nor in outline. Indeed, their usual practice is to *postulate the second law of thermodynamics in terms of entropy.* Because a postulate must, as a matter of practice, be accepted, such postulation eliminates the need for proof.

However, the success of postulating the second law in terms of entropy depends on adopting a definition of entropy that incorporates its classically determined properties—properties that Clausius, building on the deep foundations of the first law and an independent version of the second law, proved are necessary.

9.4 Clausius's Theorem Illustrated

The steps of a Carnot cycle, with half of them reversed, illustrate the claim of Clausius's proof. In particular, as a system proceeds reversibly along two different paths between the same initial and final states, the entropy added to or subtracted from the system is the same.

Consider, then, two identical systems, each at temperature T_H and in the same state. The first system, while in thermal contact with the heat reservoir at temperature T_H, expands and reversibly pushes against a piston head. Suppose that during this process, the system receives heat Q_H and consequently increases its entropy by Q_H/T_H. Then, while thermally isolated, the system continues to do reversible work against the piston head until its temperature drops to T_C. These steps are illustrated in figure 9.1.

The second of the two identical systems goes through a process that switches the order of these steps. Again, its initial temperature is T_H. Then while thermally isolated, this system pushes reversibly against a piston head until its temperature drops to T_C. The system is placed in thermal contact with a heat reservoir at temperature T_C and then reversibly expands until it receives heat Q_C that is, by design, equal to $T_C \cdot Q_H/T_H$. Consequently, the system increases its entropy by Q_C/T_C or by its equivalent Q_H/T_H. These steps are illustrated in figure 9.2.

In this way, two identical systems identically initiated travel by different reversible paths to the same final state. As expected, the entropy of the first system increases by Q_H/T_H, while the entropy of the second increases by the same increment, Q_C/T_C [$= Q_H/T_H$]. Yet these processes are only two of an indefinitely large number of reversible paths that connect the same two

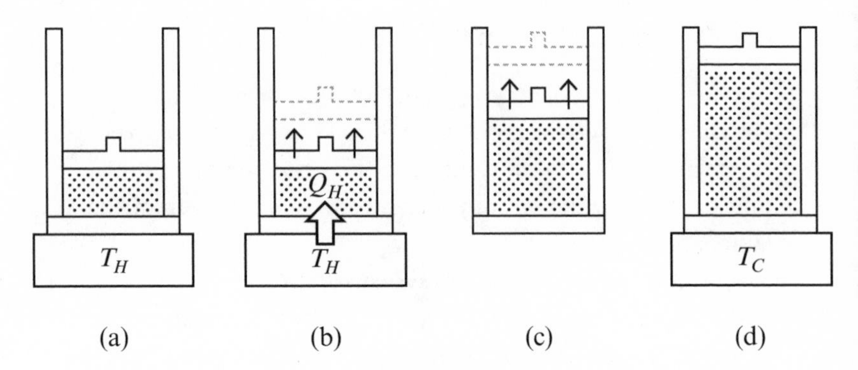

Figure 9.1
(a) The system is in thermal contact with a heat reservoir at temperature T_H. (b) While in thermal contact with the heat reservoir at temperature T_H, the system pushes reversibly against a piston and absorbs heat Q_H. (c) While thermally isolated, the system pushes reversibly against the piston until its temperature drops from T_H to T_C. (d) The system is in thermal contact with a heat reservoir at temperature T_C.

states of identical systems and, according to Clausius's theorem, increment these systems by the same amount.

9.5 But What Is Entropy, Really?

The laws of classical thermodynamics tell us only that the state variable entropy exists and the entropy of a system with temperature T changes by $\delta Q/T$ when it reversibly receives or rejects the signed quantity of heat δQ. Therefore, to ask, "But what is entropy, really?" is to ask a question to which classical thermodynamics has no answer.

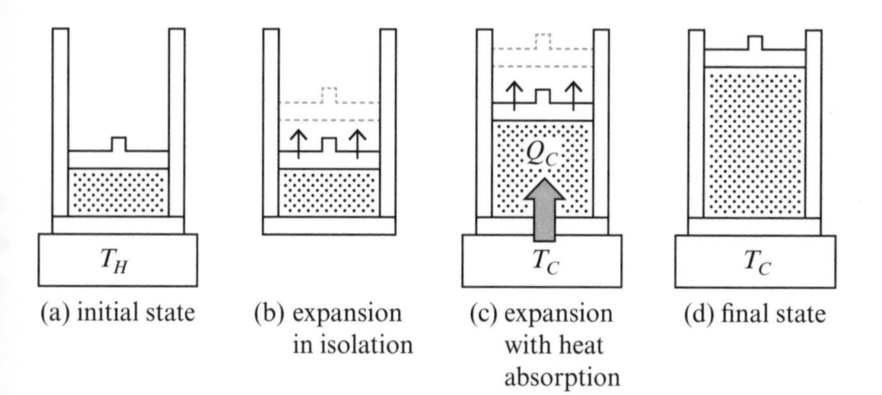

Figure 9.2

(a) The system is in thermal contact with a heat reservoir at temperature T_H. (b) While thermally isolated, the system pushes reversibly against a piston until its temperature drops to T_C. (c) While in thermal contact with the heat reservoir at temperature T_C, the system pushes reversibly against the piston until it absorbs heat Q_C that is, by design, equal to $T_C \cdot Q_H / T_H$. (d) The system is in thermal contact with a heat reservoir at temperature T_C.

The laws of classical thermodynamics say just as little about the state variable *energy*. In particular, the first law says only that the state variable *energy* exists and is changed by the processes of heating, cooling, working, and being worked on—and nothing else.[4] Our relative familiarity with the concept of energy derives solely from its additional application to Newtonian systems.

That classical thermodynamics is so sparing in its descriptions of important concepts is a sign of its durability and versatility.

4. True except when the identity of the system is changed when mass is transferred to or from the system.

Classical thermodynamics is durable because it has remained unchanged as physics changed dramatically in the twentieth century. Classical thermodynamics is versatile because, just as thermodynamic energy has been adapted to and identified in Newtonian systems, thermodynamic entropy has been identified with statistical entropy. Indeed, one expects classical thermodynamics to survive yet new developments and for its concepts to be yet newly adapted.

Here Clausius describes how the "equivalence value" of a system is changed under reversible and irreversible cyclic transformations. (In 1862, he had not yet coined the word entropy.*) As such, this excerpt references the physics not only of this chapter and but also of chapter 10.—DSL*

Rudolf Clausius, 1862

"Sixth Memoir: On the Application of the Theorem of the Equivalence of Transformations to Interior Work,"* in his *The Mathematical Theory of Heat* (London: John Van Voorst, 1867), 215–219.

In a memoir published in the year 1854,† wherein I sought to simplify to some extent the form of the developments I had previously published, I deduced, from my fundamental proposition

*Communicated to the Naturforschende Gesellschaft of Zurich, Jan. 27th, 1862; published in the Vierteljahrschrift of this Society, vol. vii. p. 48; in *Poggendorff's Annalen,* May 1862, vol. cxvi. p. 73; in the *Philosophical Magazine,* S. 4. vol. xxiv. pp. 81, 201; and in the *Journal des Mathematiques of Paris,* S. 2. vol. vii. p. 209.

†"On a modified form of the second Fundamental Theorem in the Mechanical Theory of heat." [Fourth Memoir of this collection, p. 116.]

that heat cannot, by itself, pass from a colder into a warmer body, a theorem which is closely allied to, but does not entirely coincide with, the one first deduced by S. Carnot from considerations of a different kind, based upon the older views of the nature of heat. It has reference to the circumstances under which work can be transformed into heat, and conversely, heat converted into work; and I have called it the *Theorem of the Equivalence of Transformations*. I did not, however, there communicate the entire theorem in the general form in which I had deduced it, but confined myself on that occasion to the publication of a part which can be treated separately from the rest, and is capable of more strict proof.

In general, when a body changes its state, work is performed *externally* and *internally* at the same time—the exterior work having reference to the forces which extraneous bodies exert upon the body under consideration, and the interior work to the forces exerted by the constituent molecules of the body in question upon each other. The interior work is for the most part so little known, and connected with another equally unknown quantity[‡] in such a way, that in treating of it we are obliged in some measure to trust to probabilities; whereas the exterior work is immediately accessible to observation and measurement, and thus admits of more strict treatment. Accordingly, since, in my former paper, I wished to avoid everything that was hypothetical, I entirely excluded the interior work, which I was able to do by confining myself to the consideration of *cyclical processes*—that is to say, operations in which the modifications which the body undergoes are so arranged that the body finally returns to its original condition. In such operations the interior work which is performed during the several modifications, partly in a positive sense and partly in a negative sense, neutralizes itself, so that nothing but exterior work remains, for which the theorem in question can then be demonstrated with mathematical strictness, starting from the above-mentioned fundamental proposition.

[‡][In fact—with the increase of the heat actually present in the body.—1864.]

I have delayed till now the publication of the remainder of my theorem, because it leads to a consequence which is considerably at variance with the ideas hitherto generally entertained of the heat contained in bodies, and I therefore thought it desirable to make still further trial of it. But as I have become more and more convinced in the course of years that we must not attach too great weight to such ideas, which in part are founded more upon usage than upon a scientific basis, I feel that I ought to hesitate no longer, but to submit to the scientific public the theorem of the equivalence of transformations in its complete form, with the theorems which attach themselves to it. I venture to hope that the importance which these theorems, supposing them to be true, possess in connexion with the theory of heat will be thought to justify their publication in their present hypothetical form.

I will, however, at once distinctly observe that, whatever hesitation may be felt in admitting the truth of the following theorems, the conclusions arrived at in my former paper, in reference to cyclical processes, are not at all impaired.

1. I will begin by briefly stating the theorem of the equivalence of transformations, as I have already developed it, in order to be able to connect with it the following considerations.

When a body goes through a cyclical process, a certain amount of exterior work may be produced, in which case a certain quantity of heat must be simultaneously expended; or, conversely, work may be expended and a corresponding quantity of heat may be gained. This may be expressed by saying: *Heat can be transformed into work, or work into heat, by a cyclical process.* There may also be another effect of a cyclical process: heat may be transferred from one body to another, by the body which is undergoing modification absorbing heat from the one body and giving it out again to the other. In this case the bodies between which the transfer of heat takes place are to be viewed merely as heat reservoirs, of which we are not concerned to know anything except the temperatures. If the temperatures of the two bodies differ, heat passes either from a warmer to a colder body or from a colder to a warmer body, according to the direction in which the transference of heat takes place. Such a transfer of heat may also be designated, for the sake of uniformity, a *transformation,* inasmuch as it may be said that *heat of one temperature is transformed into heat of another temperature.*

The two kinds of transformations that have been mentioned are related in such a way that one presupposes the other, and that they can mutually replace each other. If we call transformations which can replace each other *equivalent,* and seek the mathematical expressions which determine the amount of the transformations in such a manner that equivalent transformations become equal in magnitude, we arrive at the following expression: *If the quantity of heat Q of the temperature t is produced from work, the equivalence-value of this transformation is*

$$\frac{Q}{T};$$

and if the quantity of heat Q passes from a body whose temperature is t_1, into another whose temperature is t_2, the equivalence-value of this transformation is

$$Q\left(\frac{1}{T_2}-\frac{1}{T_1}\right),$$

where T is a function of the temperature which is independent of the kind of process by means of which the transformation is effected, and T_1 and T_2 denote the values of this function which correspond to the temperatures t_1 and t_2. I have shown by separate considerations that T is in all probability nothing more than the *absolute temperature.*

These two expressions further enable us to recognize the positive or negative sense of the transformations. In the first, Q is taken as positive when work is transformed into heat, and as negative when heat is transformed into work. In the second, we may always take Q as positive, since the opposite senses of the transformations are indicated by the possibility of the difference

$1/T_2 - 1/T_1$ being either positive or negative. It will thus be seen that the passage of heat from a higher to a lower temperature is to be looked upon as a positive transformation, and its passage from a lower to a higher temperature as a negative transformation.

If we represent the transformations which occur in a cyclical process by these expressions, the relation existing between them can be stated in a simple and definite manner. If the cyclical process is *reversible,* the transformations which occur therein must be partly positive and partly negative, and the equivalence values of the positive transformations must be together equal to those of the negative transformations, so that the algebraic sum of all the equivalence-values becomes = 0. If the cyclical process is not reversible, the equivalence-values of the positive and negative transformations are not necessarily equal, but they can only differ in such a way that the positive transformations predominate. The theorem respecting the equivalence-values of the transformations may accordingly be stated thus:—*The algebraic sum of all the transformations occurring in a cyclical process can only be positive, or, as an extreme case, equal to nothing.*

The mathematical expression for this theorem is as follows. Let dQ be an element of the heat given up by the body to any reservoir of heat during its own changes (heat which it may absorb from a reservoir being here reckoned as negative), and T the absolute temperature of the body at the moment of giving up this heat, then the equation

$$\int \frac{dQ}{T} = 0$$

must be true for every reversible cyclical process, and the relation

$$\int \frac{dQ}{T} \geq 0 \tag{Ia}$$

must hold good for every cyclical process which is in any way possible.

10
Law of Entropy Nondecrease

10.1 Irreversible Heat Transfer

The simplest instance of *irreversible* heat transfer occurs when a hotter reservoir transfers heat Q *directly* to a colder reservoir as illustrated in figure 10.1. In this case, we would like to know how much entropy is gained and lost by each reservoir. Yet the method justified by Clausius's theorem for keeping account of entropy loss and gain in terms of the increments Q / T applies only when the process is reversible.

Recall, though, that entropy is a state variable and that every state of every thermodynamic system has an entropy relative to the conventional entropy of some conveniently chosen state. And since there is a reversible path connecting any two states, one can calculate the net entropy increment of a system that proceeds along one of those reversible paths. Therefore, the entropy increment occasioned by an irreversible process connecting any two states is identical to that caused by a reversible process connecting the same two states.

The following comparison may help clarify this idea. Imagine a country of contiguous territory in which every location has an elevation with respect to some arbitrarily chosen level. Unfortunately, the altimeter available to its citizens works accurately

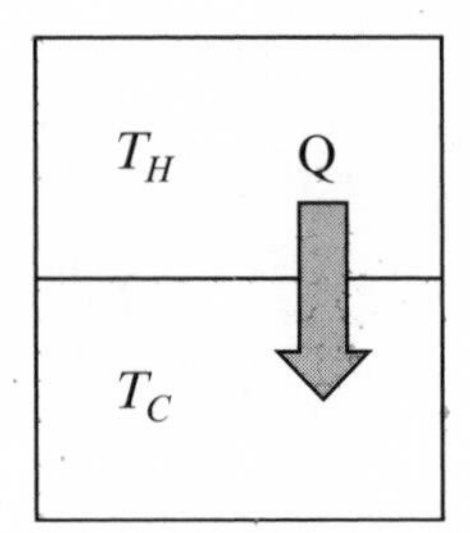

Figure 10.1
Irreversible heat flow between a hotter reservoir at temperature T_H and a colder one at temperature Q.

only when it is carried by surface route, not when taken aloft. Thus, an explorer who for the first time visits a relatively inaccessible location via balloon (that is, via an irreversible process) does not know that location's altitude (that is, its entropy). Yet the explorer realizes that this location has an altitude that in principle can be measured via surface route (that is, via a reversible process).

In similar fashion, we can devise a reversible process that achieves the same result as an irreversible one—in particular, as the direct, irreversible heat transfer between the two heat reservoirs of figure 10.1. Figure 10.2 illustrates such a reversible process. In step a, an expanding fluid does reversible work on a piston while reversibly receiving heat Q from the hotter heat reservoir, at temperature T_H, with which it is in thermal contact. Therefore, in this step, the hotter reservoir transfers entropy Q / T_H to the fluid. In step b, the fluid is thermally insulated and continues to do reversible work on the piston until its temperature drops to T_C. Finally, in step c, the piston comes into thermal

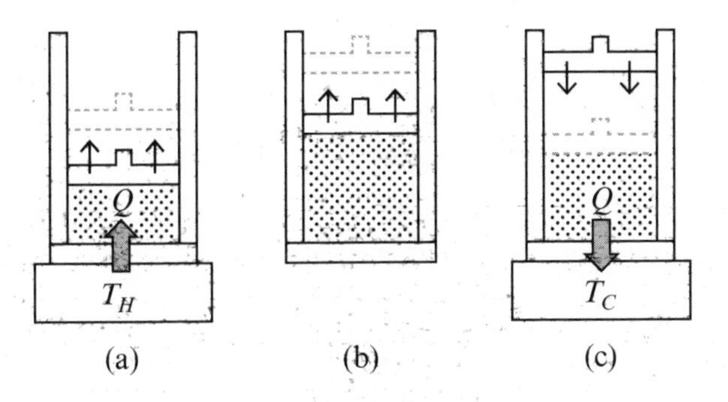

Figure 10.2
Reversible processes that take heat Q from the hotter reservoir at temperature T_H and inject heat Q into the colder reservoir at temperature T_C. The fluid does not return to its initial state.

contact with the colder reservoir at temperature T_C and reversibly compresses the fluid while transferring heat Q to the colder reservoir. In this step, the colder reservoir receives entropy Q / T_C from the system. Consequently, the entropy increment of the two reservoirs conceived as a single system is given by

$$\Delta S = Q\left(-\frac{1}{T_H} + \frac{1}{T_C}\right), \tag{10.1}$$

and because $T_C < T_H$, $\Delta S > 0$. Therefore, the net entropy of the two reservoirs increases. Note that in this case, the fluid does not (and need not) return to its initial state.

10.2 Irreversible Work

Direct heat transfer is not the only means of irreversibly changing the entropy of a system. Figure 10.3, consisting of a Joule

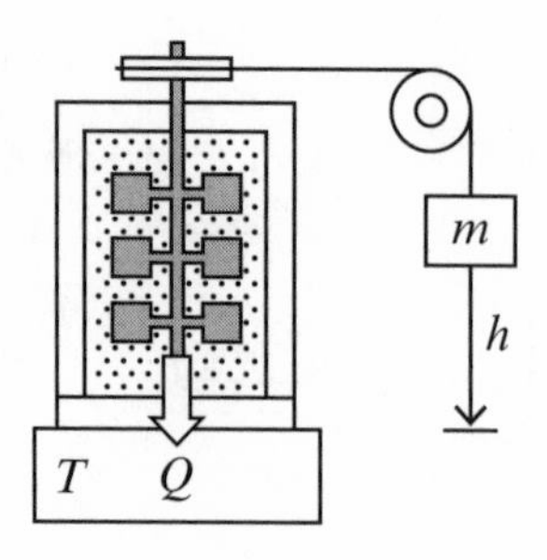

Figure 10.3
Joule apparatus in thermal contact with a heat reservoir.

apparatus in thermal contact with a heat reservoir, illustrates this point.

The Joule apparatus contains a fluid, initially at temperature T, thermally insulated except at one side in thermal contact with a heat reservoir also at temperature T. A falling mass m pulls a cord that runs over a frictionless pulley and turns a paddlewheel immersed in the fluid. As the mass m descends a distance h, the earth-mass system does work mgh on the fluid—irreversible work since the paddle sets up convection currents whose energy is dissipated in the fluid. Thus, while being stirred, the fluid has no unique temperature. After the stirring is complete, the fluid temperature returns to its original value T. In the process, heat Q, equal in magnitude to mgh, is irreversibly transferred to the reservoir.

We need no special imagination to conceive of a reversible process that achieves the same result. That illustrated in figure 10.2c in which a piston does reversible work on a fluid in thermal equilibrium with a heat reservoir suffices. Since in this process, heat Q is reversibly transferred to the heat reservoir at a temperature T, the entropy of the reservoir increases by Q/T. Therefore, the irreversible process of the Joule apparatus also increases the entropy of the heat reservoir with which it is in contact by $Q/T[= mgh/T]$.

10.3 The Law of Entropy Nondecrease

The reversible interactions described in chapter 9 teach us that the entropy of an isolated, reversibly transforming system is conserved. Those of sections 10.1 and 10.2 teach us that the entropy of an isolated, irreversibly transforming system increases. In short, *the entropy of an isolated system never decreases.*

The word *isolated,* as used here, requires comment. It is, of course, possible to isolate a thermodynamic system with perfectly insulating boundaries and rigid, impermeable walls. But any nonisolated system can also be made part of a larger isolated one by including within the larger system all parts of the universe with which the smaller system interacts, and all parts that interact with those parts, and so on.

Only processes that observe both the first and an independent version of the second law of thermodynamics observe the law of entropy nondecrease. After all, Clausius could define an entropy increment only for processes in which both laws were observed. Therefore, processes that do not observe both the first and an independent version of the second law of thermodynamics violate the law of entropy nondecrease.

Consider, for instance, a process that simply transfers a quantity of heat Q from a colder reservoir at temperature T_C to a hotter one at T_H, in this way violating the Clausius version of the second law. The colder reservoir loses entropy Q/T_C, while the hotter one gains entropy Q/T_H. The net change in the entropy ΔS of the system composed of both reservoirs is then given by

$$\Delta S = Q\left(\frac{1}{T_H} - \frac{1}{T_C}\right) < 0 \ . \qquad (10.2)$$

In this case, the first law is observed, the Clausius version of the second law is violated, and the entropy of the isolated system composed of both heat reservoirs decreases.

Similarly, a process that merely extracts heat Q from a reservoir at temperature T and produces an equivalent amount of work $W[=Q]$, in this way violating the Thomson version of the second law, also violates the law of entropy nondecrease. In this case, the isolated system composed of the heat reservoir and the mechanism producing work loses entropy Q/T. Thus, when the first law is observed and the Thomson version of the second law is violated, the entropy of an isolated system decreases.

Processes that observe an independent version of the second law yet violate the first law of thermodynamics also violate the law of entropy nondecrease—in a yet more fundamental way. For instance, a process that causes a hotter heat reservoir to give up heat Q_H and a colder one to receive a different quantity of heat $Q_C[\neq Q_H]$ observes the second law yet violates the first. Yet such a processes also violates the law of entropy nondecrease. For only when the first law is observed is energy is a state variable. And if energy is not a state variable, then Thomson's absolute temperature of 1851 could not be defined. And if Thomson's absolute temperature of 1851 cannot be defined, then entropy is not a state variable. And if entropy is not a state variable, no more can be said of "the entropy of a system" than can be said of the "work of a system" or the "heat of a system." It is in this way that the law of entropy nondecrease depends on the first law of thermodynamics.

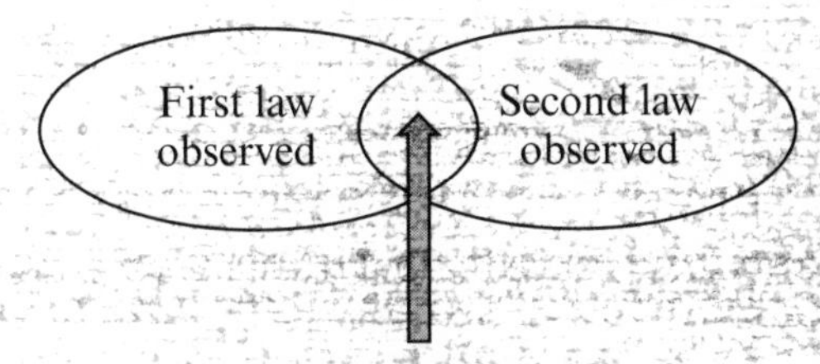

Figure 10.4
Relation among processes in which the first and second laws of thermodynamics are observed.

Figure 10.4 outlines how the first law of thermodynamics and an independent version of the second law are related to the law of entropy nondecrease. The two ovals encompass putative processes in which, respectively, the first and an independent version of the second law are observed, while their intersection encompasses processes in which both laws of classical thermodynamics are observed. It is only in the intersection of the two ovals that the state variable entropy exists and the law of entropy nondecrease is observed. Processes that observe one but not both of the two laws of classical thermodynamics violate the law of entropy nondecrease.

11
The Fate of the Universe?

11.1 Energy Dissipation

Rudolf Clausius and William Thomson had been speaking figuratively of entropy and its consequences for over a decade before Clausius fashioned the word *entropy* out of Greek stems in 1865. The title of Thomson's 1852 article, "On the Universal Tendency in Nature to the Dissipation of Mechanical Energy," established Thomson's preferred language.[1] According to Thomson, *entropy increases when mechanical energy dissipates*, and dissipated mechanical energy loses its ability to do work.

Yet however intimate their relation, there is no perfect equation between entropy increase and the dissipation of mechanical energy, for the phrase *dissipated mechanical energy* does not capture all aspects of entropy increase. Consider, for instance, two equal-temperature, equal-pressure gases with different compositions expanding into and mixing with each other, as illustrated in figure 11.1. As each gas expands without doing work, its temperature remains constant. Even so, each gas increases its entropy. (Recall the ideal gas in figure 10.2a that increases its entropy while maintaining its temperature and

1. *Mathematical and Physical Papers*, article 59, 511–514, and in *Proceedings of the Royal Society of Edinburgh*, April 19, 1852.

Clausius's statement on the fate of the universe is here placed in its context.—DSL

Rudolf Clausius, 1865

"Ninth Memoir: On Several Convenient Forms of the Fundamental Equations of the Mechanical Theory of Heat,"* in his *The Mathematical Theory of Heat* (London: John Van Voorst, 1867), 364–365.

17. In conclusion I wish to allude to a subject whose complete treatment could certainly not take place here, the expositions necessary for that purpose being of too wide a range, but relative to which even a brief statement may not be without interest, inasmuch as it will help to show the general importance of the magnitudes which I have introduced when formulizing the second fundamental theorem of the mechanical theory of heat.

The second fundamental theorem, in the form which I have given to it, asserts that all transformations occurring in nature may take place in a certain direction, which I have assumed as positive, by themselves, that is, without compensation; but that in the opposite, and consequently negative direction, they can only take place in such a manner as to be compensated by simultaneously occurring positive transformations. The application of this theorem to the Universe leads to a conclusion to which W. Thomson first drew attention,† and of which I have spoken in the Eighth Memoir. In fact, if in all the changes of condition occurring in the universe the transformations in one definite direction exceed in magnitude those in the opposite

*Read at the Philosophical Society of Zurich on the 24th of April, 1866, published in the Vierteljahrsschrift of this Society, Bd. x. S. 1.; Pogg. Ann., July, 1805, Bd. cxxv. S. 353 ; Journ. de Liouville, 2e ser. t. x. p. 361.
†Phil. Mag. Ser. 4. vol. iv. p. 304.

direction, the entire condition of the universe must always continue to change in that first direction, and the universe must consequently approach incessantly a limiting condition.

The question is, how simply and at the same time definitely to characterize this limiting condition. This can be done by considering, as I have done, transformations as mathematical quantities whose equivalence-values may be calculated, and by algebraical addition united in one sum.

In my former Memoirs I have performed such calculations relative to the heat present in bodies, and to the arrangement of the particles of the body. For every body two magnitudes have thereby presented themselves—the transformation-value of its thermal content, and its disgregation; the sum of which constitutes its entropy. But with this the matter is not exhausted; radiant heat must also be considered, in other words, the heat distributed in space in the form of advancing oscillations of the aether must be studied, and further, our researches must be extended to motions which cannot be included in the term Heat.

The treatment of the last might soon be completed, at least so far as relates to the motions of ponderable masses, since allied considerations lead us to the following conclusion. When a mass which is so great that an atom in comparison with it may be considered as infinitely small, moves as a whole, the transformation-value of its motion must also be regarded as infinitesimal when compared with its *vis viva*; whence it follows that if such a motion by any passive resistance becomes converted into heat, the equivalence-value of the uncompensated transformation thereby occurring will be represented simply by the transformation-value of the heat generated. Radiant heat, on the contrary, cannot be so briefly treated, since it requires certain special considerations in order to be able to state how its transformation-value is to be determined. Although I have already, in the Eighth Memoir above referred to, spoken of radiant heat in connexion with the mechanical theory of heat, I have not alluded to the present question, my sole intention being to prove that no contradiction exists between the laws of radiant heat and an axiom assumed by

12
Classical Thermodynamics

12.1 The Third Law

Walter Nernst (1864–1941) was the first, in 1906, to gather evidence, based on the heat produced by chemical reactions at low temperatures, for the existence of a new *heat theorem*, now called the *third law of thermodynamics*.[1] But it was Max Planck (1858–1947) who gave Nernst's theorem the following form:[2]

> As the temperature diminishes indefinitely the entropy of a chemical homogeneous body of finite density approaches indefinitely near to a finite value, which is independent of the pressure, the state of aggregation and of the special chemical modification. (1917)

Planck also suggested that the finite value, independent of all thermodynamic state variables, to which the entropy of a system approaches as its absolute temperature approaches zero could in

1. See Walter Nernst, *The New Heat Theorem* (New York: Dover, 1969), 1–4 (first published in English in 1926).
2. Max Planck, *Treatise on Thermodynamics*, 3rd ed. (New York: Dover, n.d.), 273 (first published in English in 1917). See Ralph Baierlein, *Thermal Physics* (Cambridge: Cambridge University Press, 1999), 331–347, for various forms of the third law of thermodynamics.

each case be set equal to a common number conveniently chosen to be zero—a choice I call *Planck's convention.*

Thus, the third law constrains only the low-temperature behavior of thermodynamic systems. If a system's proposed equations of state violate either the first or second laws, they are worthless. If they violate the third law, as do, for instance, the ideal gas equations of state, their use is merely limited to sufficiently high temperatures.

"The designation *third law,* although very common, is in some ways pretentious. The third law is neither as foundational nor as consequential as the other laws of thermodynamics. Yet it cannot be derived from the other laws and it provides a useful constraint on low-temperature experiments and theories."[3]

12.2 The Zeroth Law

Nowhere in our ordering are chronology and logic more at odds than in the position assigned the zeroth law of thermodynamics, for the zeroth law underlies the concept of empirical temperature. And without empirical temperature, a unit of heat could not be defined or the second law of thermodynamics be expressed. And without carefully defining a unit of heat, the first law of thermodynamics would not have been discovered. The first law, in turn, leads to the state variable energy, and without the first and second laws, there is no state variable entropy. And, of course, the third law constrains the low-temperature limit of entropy. In this way, the zeroth law underlies all of classical thermodynamics.

3. Don S. Lemons, *Mere Thermodynamics* (Baltimore: Johns Hopkins University Press, 2009), 164.

Indeed, all those who from the eighteenth century on have thought carefully about the meaning of empirical temperature have assumed, however implicitly, the zeroth law. But it was not until the 1930s, in an effort to establish thermodynamics on a set of axioms, that the zeroth law was identified as such.[4]

The zeroth law affirms that *two systems, each in thermal equilibrium with a third, are in thermal equilibrium with each other*. In a word, the relation of thermal equilibrium is *transitive*.

Thermal equilibrium itself refers to the stasis achieved when two bodies are in thermal contact, that is, when allowed to interact through a boundary that allows heat flow but prohibits work and mass exchange. Hot coffee held within a ceramic cup achieves thermal equilibrium with its room-temperature environment within about an hour.

The zeroth law authorizes the concept of the thermometer, for the thermometric state variable of a thermometer is simply an indicator of thermal equilibrium. If a thermometer is brought into successive thermal equilibrium with two different systems and its thermometric state variable does not change, then the two systems are in thermal equilibrium with each other, as well as with the thermometer. Therefore, these two systems have the same temperature.

12.3 Einstein on Classical Thermodynamics

The four laws of classical thermodynamics can be summarized as follows:

4. According to Arnold Sommerfeld, it was Ralph H. Fowler who invented the term *zeroth law of thermodynamics* around 1935. See Arnold Sommerfeld, *Thermodynamics and Statistical Mechanics*, vol. 1, *Lectures on Theoretical Physics* (Cambridge, MA: Academic Press, 1951/1955), 1.

the world, however limited in perspective, is, as Einstein seemed to imply, of permanent value? Of classical thermodynamics, he had no doubt. Within its framework of applicability, classical thermodynamics is "the only physical theory of universal content" that "will never be overthrown."

Annotated Bibliography

A book appears here because I have consulted it or cited it frequently, I recommend it for further study, or its approach contrasts revealingly with that of *Thermodynamic Weirdness*. When two publication dates appear, the first is that of the edition or reprint I consulted and the second that of the original edition.

Baierlein, Ralph. *Thermal Physics*. Cambridge: Cambridge University Press, 1999. This book and the one listed below by Daniel Schroeder are skillfully executed examples of the thermal physics approach to thermodynamics and statistical mechanics. 460 pages.

Bridgman, Percy. *The Nature of Thermodynamics*. New York: Harper, 1961, 1941. A rambling but sometimes insightful analysis of classical thermodynamics by a Nobel prize–winning physicist. 239 pages.

Caneva, Kenneth L. *Robert Mayer and the Conservation of Energy*. Princeton: Princeton University Press, 1993. A careful historical analysis. 468 pages.

Caneva, Kenneth L. "Colding, Ørsted, and the Meanings of Force." *Historical Studies in the Physical and Biological Sciences* 28, no. 1 (1997). The meanings of the word *force*, as used in this scholarly investigation, encompass what is now meant by *energy*. 138 pages.

Cardwell, D. S. L. *From Watt to Clausius: The Rise of Thermodynamics in the Early Industrial Age.* Ithaca, NY: Cornell University Press, 1971. An excellent, expansive history of thermodynamics that, unfortunately, is out of print. 336 pages.

Carnot, Sadi. *Reflections on the Motive Power of Fire and Other Papers on the Second Law of Thermodynamics.* Gloucester, MA: Peter Smith, 1977, 1960. This book contains an introduction by the historian E. Mendoza and papers by Sadi Carnot, Emile Clapeyron, and Rudolf Clausius, as well as selections from Carnot's posthumous notes. The first twenty-two pages of Carnot's seventy-page *Reflections on the Motive Power of Fire* contain his version of the second law of thermodynamics and a proof of what has become known as Carnot's theorem. Also available as an inexpensive Dover Publications reprint. 152 pages.

Chang, Hasok. *Inventing Temperature: Measurement and Scientific Progress.* Oxford: Oxford University Press, 2004. A historical and philosophical study of the concept of temperature. Chang, in the words of one reviewer, "conveys beautifully the feeling of what it is like not to know something before it is figured out." 304 pages.

Clausius, Rudolf. *The Mechanical Theory of Heat.* London: John Van Voorst, 1867. A collection of Clausius's nine papers on thermodynamics, including those in which he identifies the first and second laws of thermodynamics and the concept of entropy. 374 pages.

Lemons, Don S. *Mere Thermodynamics.* Baltimore: Johns Hopkins University Press, 2009. The structure and content of this mathematically oriented textbook anticipate the approach in *Thermodynamic Weirdness.* 224 pages.

Magie, W. F. *A Source Book in Physics.* Cambridge, MA: Harvard University Press, 1963, 1935. An encyclopedic anthology of excerpts from primary sources in physics. 636 pages.

Mott-Smith, Morton. *The Concept of Energy, Simply Explained.* New York: Dover, 1964, 1934. For decades, this book was the only history of classical thermodynamics available. It is still one of the best. Mott-Smith carefully blends scientific, historical, and philosophical material and makes significant use of primary sources. 223 pages.

Muller, Ingo. *A History of Thermodynamics: The Doctrine of Energy and Entropy*. New York: Springer, 2007. A detailed, if idiosyncratic, history of thermodynamics. 320 pages.

Planck, Max. *Treatise on Thermodynamics*. New York: Dover, 1969, 1917. Planck discusses the third law of thermodynamics in the last few pages of this book. 320 pages.

Schroeder, Daniel. *An Introduction to Thermal Physics*. London: Pearson, 1999. See the comment on Baierlein's book, already cited. 422 pages.

Thomson, William. *Mathematical and Physical Papers*, vol. 1. Cambridge: Cambridge University Press, 1882. Thompson's contributions to thermodynamics appear in this volume of his collected papers. 727 pages.

Truesdell, C. *The Tragicomical History of Thermodynamics, 1822–1854*. New York: Springer, 1980. A history of thermodynamics that gives priority to its mathematical formulation. 372 pages.

Vanderslice, J. T., H. W. Schamp Jr., and E. A. Mason. *Thermodynamics*. Englewood Cliffs, NJ: Prentice Hall, 1966. Little known and long out of print, this textbook was the first to emphasize that the second law of thermodynamics has consequences of its own (including those derived by Carnot) that are independent of the first law. 244 pages.

Von Baeyer, Hans Christian. *Warmth Disperses and Time Passes: The History of Heat*. New York: Modern Library, 1999. A well-written and well-researched popular history of thermodynamics. 240 pages.

Index